MEI STRUCTURED MATHEMATICS

SECOND EDITION

Statistics 3

Anthony Eccles
Nigel Green
Roger Porkess

Series Editor: Roger Porkess

Hodder & Stoughton

A MEMBER OF THE HODDER HEADLINE GROUP

Acknowledgements

We are grateful to the following companies, institutions and individuals who have given permission to reproduce photographs in this book. Every effort has been made to trace and acknowledge ownership of copyright. The publishers will be glad to make suitable arrangements with any copyright holders whom it has not been possible to contact.

James L. Amos/CORBIS (page 1); Helen Litt (page 4); Kevin R. Morris/CORBIS (page 34); Hodder Picture Library (pages 53 and 74); Matt Fullerty (pages 94, 102 and 117).

OCR, AQA and Edexcel accept no responsibility whatsoever for the accuracy or method of working in the answers given.

Orders: please contact Bookpoint Ltd, 78 Milton Park, Abingdon, Oxon OX14 4TD. Telephone: (44) 01235 827720, Fax: (44) 01235 400454. Lines are open from 9.00–6.00, Monday to Saturday, with a 24 hour message answering service. Email address: orders@bookpoint.co.uk

British Library Cataloguing in Publication Data
A catalogue record for this title is available from The British Library

ISBN 0340 77 1992

First published 1993
Second edition 2000
Impression number   10 9 8 7 6 5 4 3 2 1
Year   2005 2004 2003 2002 2001 2000

Copyright © 1993, 2000 Anthony Eccles, Nigel Green, Roger Porkess

Typeset by Tech-Set Ltd, Gateshead, Tyne & Wear.
Printed in Great Britain for Hodder & Stoughton Educational, a division of Hodder Headline Plc, 338 Euston Road, London NW1 3BH by Martins the printers, Berwick upon Tweed.

MEI Structured Mathematics

Mathematics is not only a beautiful and exciting subject in its own right but also one that underpins many other branches of learning. It is consequently fundamental to the success of a modern economy.

MEI Structured Mathematics is designed to increase substantially the number of people taking the subject post-GCSE, by making it accessible, interesting and relevant to a wide range of students.

It is a credit accumulation scheme based on 45 hour modules which may be taken individually or aggregated to give Advanced Subsidiary (AS) and Advanced GCE (A Level) qualifications in Mathematics, Further Mathematics and related subjects (like Statistics). The modules may also be used to obtain credit towards other types of qualification.

The course is examined by OCR (previously the Oxford and Cambridge Schools Examination Board) with examinations held in January and June each year.

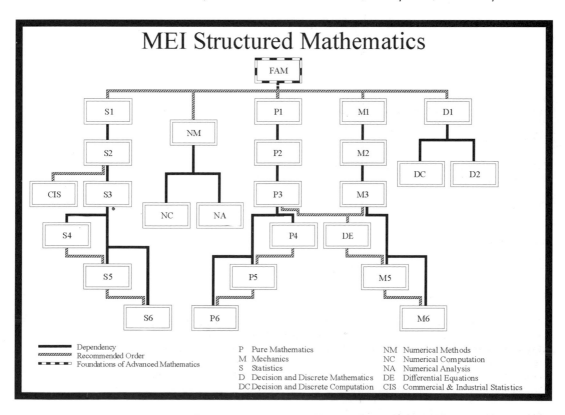

This is one of the series of books written to support the course. Its position within the whole scheme can be seen in the diagram above.

Mathematics in Education and Industry is a curriculum development body which aims to promote the links between Education and Industry in Mathematics at secondary level, and to produce relevant examination and teaching syllabuses and support material. Since its foundation in the 1960s, MEI has provided syllabuses for GCSE (or O Level), Additional Mathematics and A Level.

For more information about MEI Structured Mathematics or other syllabuses and materials, write to MEI Office, Albion House, Market Place, Westbury, Wiltshire, BA13 3DE.

Introduction

This is the third in a series of books to support the Statistics modules in MEI Structured Mathematics. The first three together cover the statistics requirements of somebody taking an A level in Pure Mathematics and Statistics and, as such, are suitable for use on a variety of courses.

In all branches of MEI Structured Mathematics the emphasis is on understanding, interpretation and modelling. This book begins with a chapter covering the techniques for continuous models, and this is followed by one on the sums and differences of random variables. The final three chapters are concerned with the collection and interpretation of sample data and introduce you to the use of sample data for hypothesis testing, based on the normal, t- and χ^2 distributions, and for constructing confidence intervals.

As in the two earlier statistics books, several examples are taken from the pages of a fictional local newspaper, The *Avonford Star*. Much of the information that you receive from the media is of a broadly statistical nature. In these books you are encouraged to recognise this and are shown how to evaluate what you are told.

In a number of the contexts we have chosen to use the everyday English word 'weight' in preference to the more formal 'mass'.

This is the second edition of this book and it includes more past examination questions. The first edition included a chapter on sampling methods but this is now covered in *Statistics 1*. New to this edition is the section on Type I and Type II errors.

The authors would like to thank the many people who have helped in the preparation of this book and its two predecessors, particularly Graham Cooper, Kath Fearns and Neil Sheldon who advised us on all three of the original books; for the second edition we would like to add Bill Gibson, Ruth Stanier and Mary Tappenden to this list. We would also like to thank the various examination boards who have given permission for their past questions to be included in the exercises.

Anthony Eccles, Nigel Green and Roger Porkess

Contents

Continuous random variables

A theory is a good theory if it satisfies two requirements: It must accurately describe a large class of observations on the basis of a model that contains only a few arbitrary elements, and it must make definite predictions about the results of future observations.

Stephen Hawking
A Brief History of Time

THE AVONFORD STAR

Lucky escape for local fisherman

Local fisherman George Sutherland stared death in the face yesterday as he was plucked from the deck of his 56 ft boat, the *Belle Star*, by a freak wave. Only the quick thinking of his brother James, who grabbed hold of his legs, saved George from a watery grave.

'It was a bad day and suddenly this lump of water came down on us,' said George. 'It was a wave in a million, higher than the mast of the boat, and it caught me off guard'.

Hero James is a man of few words. 'All in the day's work' was his only comment.

The Sutherland brothers contemplate the 'wave in a million' which almost caused a tragedy

Freak waves do occur and they can have serious consequences in terms of damage to shipping, oil rigs and coastal defences, sometimes resulting in loss of life. It is important to estimate how often they will occur, and how high they will be. Was George Sutherland's one in a million estimate for a wave higher than the mast of the boat (11 metres) at all accurate?

Before you can answer this question, you need to know the *probability density* of the heights of waves at that time of the year in the area where the *Belle Star* was fishing. The graph in figure 1.1 shows this sort of information; it was collected in the same season of the year as the Sutherland accident by the Offshore Weather Ship *Juliet* in the North Atlantic.

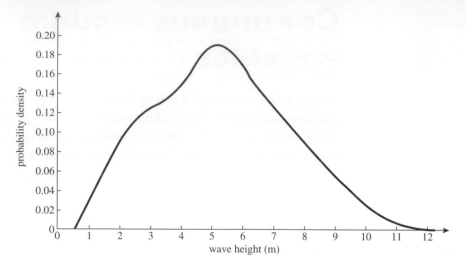

Figure 1.1

To obtain figure 1.1 a very large amount of wave data had to be collected. This allowed the class interval widths of the wave heights to be sufficiently small for the outline of the curve to acquire this shape. It also ensured that the sample data were truly representative of the population of waves at that time of the year.

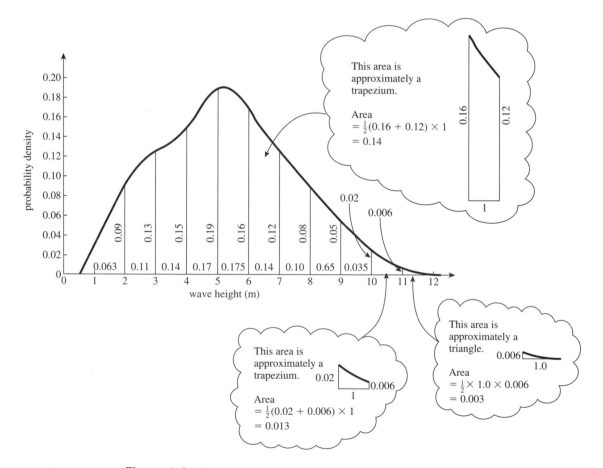

Figure 1.2

In a graph such as figure 1.1 the vertical scale is a measure of probability density. Probabilities are found by estimating the area under the curve. The total area is 1.0, meaning that effectively all waves at this place have heights between 0.6 and 12.0 m, see figure 1.2.

If this had been the place where the *Belle Star* was situated, the probability of encountering a wave at least 11 m high would have been 0.003, about 1 in 300. Clearly George's description of it as 'a wave in a million' was not justified purely by its height. The fact that he called it a 'lump of water' suggests that perhaps it may have been more remarkable for its steep sides than its height.

Probability density function

In the wave height example the curve was determined experimentally, using equipment on board the Offshore Weather Ship *Juliet*. The curve is continuous because the random variable, the wave height, is continuous and not discrete. The possible heights of waves are not restricted to particular steps (say every $\frac{1}{2}$ metre), but may take any value within a range.

 Is it reasonable to describe the height of a wave as *random*?

A function represented by a curve of this type is called a *probability density function*, often abbreviated to p.d.f. The probability density function of a continuous random variable, X, is usually denoted by $f(x)$. If $f(x)$ is a p.d.f. it follows that:

- $f(x) \geqslant 0$ for all x You cannot have negative probabilities.
- $\displaystyle\int_{\substack{\text{All} \\ \text{values} \\ \text{of } x}} f(x)\,dx = 1$ The total area under the curve is 1.

For a continuous random variable with probability density function $f(x)$, the probability that X lies in the interval $[a, b]$ is given by

$$P(a \leqslant X \leqslant b) = \int_a^b f(x)\,dx.$$

You will see that in this case the probability density function has quite a complicated curve and so it is not possible to find a simple algebraic expression with which to model it.

Most of the techniques in this chapter assume that you do in fact have a convenient algebraic expression with which to work. However, the methods are still valid if this is not the case, but you would need to use numerical, rather than algebraic, techniques for integration and differentiation. In the high-wave incident mentioned on page 1, the areas corresponding to wave heights of less than 2 m and of at least 11 m were estimated by treating the shape as a triangle: other areas were approximated by trapezia.

Note: Class boundaries

If you were to ask the question *'What is the probability of a randomly selected wave being exactly 2 m high?'* the answer would be zero. If you measured a likely looking wave to enough decimal places (assuming you could do so), you would eventually come to a figure which was not zero. The wave height might be 2.01... m or 2.000 003... m but the probability of it being exactly 2 m is infinitessimally small. Consequently in theory it makes no difference whether you describe the class interval from 2 to 2.5 m as $2 < h < 2.5$ or as $2 \leqslant h \leqslant 2.5$.

However, in practice, measurements are always rounded to some extent. The reality of measuring a wave's height means that you would probably be quite happy to record it to the nearest 0.1 m and get on with the next wave. So, in practice, measurements of 2.0 m and 2.5 m probably will be recorded, and intervals have to be defined so that it is clear which class they belong to. You would normally expect $<$ at one end of the interval and $\leqslant$ at the other: either $2 \leqslant h < 2.5$ or $2 < h \leqslant 2.5$. In either case the probability of the wave being within the interval would be given by

$$\int_2^{2.5} f(x)\, dx.$$

THE AVONFORD STAR
Rufus foils council office break-in

Somewhere an empty-pocketed thief is nursing a sore leg and regretting the loss of a pair of trousers. Council porter Fred Lamming, and Rufus, a wiry-haired Jack Russell, were doing a late-night check round the council head office when they came upon the intruder on the ground floor.

'I didn't need to say anything,' Fred told me; 'Rufus went straight for him and grabbed him by the leg.' After a tussle the man's trousers tore, leaving Rufus with a mouthful of material while the man made good his escape out of a window.

Following the incident, Avonford Council are looking at an electronic security system. 'Rufus won't live for ever,' explained Council leader Sandra Martin.

EXAMPLE 1.1

Avonford District Council are thinking of fitting an electronic security system inside head office. They have been told by manufacturers that the lifetime, X years, of the system they have in mind has the p.d.f.:

$$f(x) = \frac{3x(20 - x)}{4000} \quad \text{for } 0 \leqslant x \leqslant 20,$$

and $\qquad\qquad f(x) = 0 \qquad\qquad$ otherwise.

(i) Show that the manufacturers' statement is consistent with f(x) being a probability density function.

(ii) Find the probability that:

  (a) it fails in the first year

  (b) it lasts 10 years but then fails in the next year.

SOLUTION

(i) The condition $f(x) \geqslant 0$ for all values of x between 0 and 20 is satisfied, as shown by the graph of $f(x)$, figure 1.3.

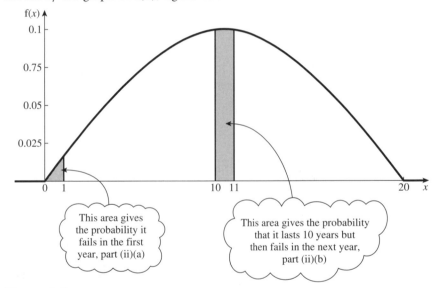

Figure 1.3

The other condition is that the area under the curve is 1.

$$\text{Area} = \int_{-\infty}^{\infty} f(x)\,dx = \int_{0}^{20} \frac{3x(20 - x)}{4000}\,dx$$

$$= \frac{3}{4000} \int_{0}^{20} (20x - x^2)\,dx$$

$$= \frac{3}{4000} \left[10x^2 - \frac{x^3}{3} \right]_{0}^{20}$$

$$= \frac{3}{4000} \left[10 \times 20^2 - \frac{20^3}{3} \right]$$

$$= 1, \text{ as required.}$$

(ii) (a) *It fails in the first year.*

This is given by $P(X < 1) = \int_0^1 \frac{3x(20 - x)}{4000}\,dx$

$$= \frac{3}{4000} \int_0^1 (20x - x^2)\,dx$$

$$= \frac{3}{4000} \left[10x^2 - \frac{x^3}{3} \right]_0^1$$

$$= \frac{3}{4000} \left(10 \times 1^2 - \frac{1^3}{3} \right)$$

$$= 0.007\,25$$

(b) *It fails in the 11th year.*

This is given by $P(10 \leqslant X < 11)$

$$= \int_{10}^{11} \frac{3x(20 - x)}{4000}\,dx$$

$$= \frac{3}{4000} \left[10x^2 - \frac{1}{3}x^3 \right]_{10}^{11}$$

$$= \frac{3}{4000} \left(10 \times 11^2 - \frac{1}{3} \times 11^3 \right) - \frac{3}{4000} \left(10 \times 10^2 - \frac{1}{3} \times 10^3 \right)$$

$$= 0.074\,75$$

EXAMPLE 1.2

The continuous random variable X represents the amount of sunshine in hours between noon and 4 pm at a skiing resort in the high season. The probability density function, $f(x)$, of X is modelled by

$$f(x) = \begin{cases} kx^2 & \text{for } 0 \leqslant x \leqslant 4 \\ 0 & \text{otherwise.} \end{cases}$$

(i) Find the value of k.
(ii) Find the probability that on a particular day in the high season there is more than two hours of sunshine between noon and 4 pm.

SOLUTION

(i) To find the value of k you must use the fact that the area under the graph of $f(x)$ is equal to 1.

$$\int_{-\infty}^{\infty} f(x)\,dx = \int_0^4 kx^2\,dx = 1$$

Therefore

$$\left[\frac{kx^3}{3} \right]_0^4 = 1$$

$$\frac{64k}{3} = 1$$

So
$$k = \frac{3}{64}$$

(ii)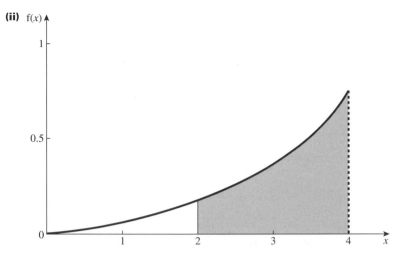

Figure 1.4

The probability of more than 2 hours of sunshine is given by

$$P(X > 2) = \int_2^\infty f(x)\,dx = \int_2^4 \frac{3x^2}{64}\,dx$$

$$= \left[\frac{x^3}{64}\right]_2^4$$

$$= \frac{64 - 8}{64}$$

$$= \frac{56}{64}$$

$$= 0.875$$

EXAMPLE 1.3

The number of hours Darren spends each day working in his garden is modelled by the continuous random variable X, with p.d.f. $f(x)$ defined by

$$f(x) = \begin{cases} kx & \text{for } 0 \leqslant x < 3 \\ k(6 - x) & \text{for } 3 \leqslant x \leqslant 6 \\ 0 & \text{otherwise.} \end{cases}$$

(i) Find the value of k.

(ii) Sketch the graph of $f(x)$.

(iii) Find the probability that Darren will work between 2 and 5 hours in his garden on a randomly selected day.

SOLUTION

(i) To find the value of k you must use the fact that the area under the graph of $f(x)$ is equal to 1. You may find the area by integration, as shown below.

$$\int_{-\infty}^{\infty} f(x)\,dx = \int_{0}^{3} kx\,dx + \int_{3}^{6} k(6-x)\,dx = 1$$

$$\left[\frac{kx^2}{2}\right]_0^3 + \left[6kx - \frac{kx^2}{2}\right]_3^6 = 1$$

Therefore $\qquad \dfrac{9k}{2} + (36k - 18k) - \left(18k - \dfrac{9k}{2}\right) = 1$

$$9k = 1$$

So $\qquad\qquad\qquad\qquad\qquad\qquad k = \tfrac{1}{9}$

Note

In this case you could have found k without integration because the graph of the p.d.f. is a triangle, with area given by $\tfrac{1}{2} \times$ base $\times$ height, resulting in the equation

$$\tfrac{1}{2} \times 6 \times k\,(6-3) = 1$$

hence $\qquad\qquad\qquad 9k = 1$

and $\qquad\qquad\qquad\quad k = \tfrac{1}{9}.$

(ii) Sketch the graph of $f(x)$.

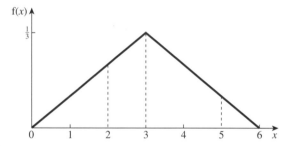

Figure 1.5

(iii) To find $P(2 \leqslant X \leqslant 5)$, you need to find both $P(2 \leqslant X < 3)$ and $P(3 \leqslant X \leqslant 5)$ because there is a different expression for each part.

$$P(2 \leqslant X \leqslant 5) = P(2 \leqslant X < 3) + P(3 \leqslant X \leqslant 5)$$

$$= \int_{2}^{3} \frac{1}{9}x\,dx + \int_{3}^{5} \frac{1}{9}(6-x)\,dx$$

$$= \left[\frac{x^2}{18}\right]_2^3 + \left[\frac{2x}{3} - \frac{x^2}{18}\right]_3^5$$

$$= \frac{9}{18} - \frac{4}{18} + \left(\frac{10}{3} - \frac{25}{18}\right) - \left(2 - \frac{1}{2}\right)$$

$$= 0.72 \text{ to two decimal places.}$$

The probability that Darren works between 2 and 5 hours in his garden on a randomly selected day is 0.72.

1 The continuous random variable X has probability density function $f(x)$ where

$$f(x) = kx \quad \text{for } 1 \leqslant x \leqslant 6$$
$$= 0 \quad \text{otherwise.}$$

(i) Find the value of the constant k.

(ii) Sketch $y = f(x)$.

(iii) Find $P(X > 5)$.

(iv) Find $P(2 \leqslant X \leqslant 3)$.

2 The continuous random variable X has p.d.f. $f(x)$ where

$$f(x) = k(5 - x) \quad \text{for } 0 \leqslant x \leqslant 4$$
$$= 0 \quad \text{otherwise.}$$

(i) Find the value of the constant k.

(ii) Sketch $y = f(x)$.

(iii) Find $P(1.5 \leqslant X \leqslant 2.3)$.

3 The continuous random variable X has p.d.f. $f(x)$ where

$$f(x) = ax^3 \quad \text{for } 0 \leqslant x \leqslant 3$$
$$= 0 \quad \text{otherwise.}$$

(i) Find the value of the constant a.

(ii) Sketch $y = f(x)$.

(iii) Find $P(X \leqslant 2)$.

4 The continuous random variable X has p.d.f. $f(x)$ where

$$f(x) = kx \quad \text{for } 0 \leqslant x \leqslant 2$$
$$= 4k - kx \quad \text{for } 2 < x \leqslant 4$$
$$= 0 \quad \text{otherwise.}$$

(i) Find the value of the constant k.

(ii) Sketch $y = f(x)$.

(iii) Find $P(1 \leqslant X \leqslant 3.5)$.

5 The continuous random variable X has p.d.f. $f(x)$ where

$$f(x) = c \quad \text{for } -3 \leqslant x \leqslant 5$$
$$= 0 \quad \text{otherwise.}$$

(i) Find c.

(ii) Sketch $y = f(x)$.

(iii) Find $P(|X| < 1)$.

(iv) Find $P(|X| > 2.5)$.

6 A continuous random variable X has p.d.f.

$$f(x) = k(x-1)(6-x) \quad \text{for } 1 \leqslant x \leqslant 6$$
$$= 0 \qquad\qquad\qquad \text{otherwise.}$$

(i) Find the value of k.

(ii) Sketch $y = f(x)$.

(iii) Find $P(2 \leqslant X \leqslant 3)$.

7 A random variable X has p.d.f.

$$f(x) = \begin{cases} (x-1)(2-x) & \text{for } 1 \leqslant x < 2 \\ a & \text{for } 2 \leqslant x \leqslant 4 \\ 0 & \text{otherwise.} \end{cases}$$

(i) Find the value of the constant a.

(ii) Sketch $y = f(x)$.

(iii) Find $P(1.5 \leqslant X \leqslant 2.5)$.

(iv) Find $P(|X - 2| < 1)$.

8 A random variable X has p.d.f.

$$f(x) = \begin{cases} kx(3-x) & \text{for } 0 \leqslant x \leqslant 3 \\ 0 & \text{otherwise.} \end{cases}$$

(i) Find the value of k.

(ii) The lifetime (in years) of an electronic component is modelled by this distribution. Two such components are fitted in a radio which will only function if both devices are working. Find the probability that the radio will still function after two years, assuming that their failures are independent.

9 The planning officer in a council needs information about how long cars stay in the car park, and asks the attendant to do a check on the times of arrival and departure of 100 cars. The attendant provides the following data:

Length of stay	Under 1 hour	1–2 hours	2–4 hours	4–10 hours	More than 10 hours
Number of cars	20	14	32	34	0

The planning officer suggests that the length of stay in hours may be modelled by the continuous random variable X with probability density function of the form

$$f(x) = \begin{cases} k(20 - 2x) & \text{for } 0 \leqslant x \leqslant 10 \\ 0 & \text{otherwise.} \end{cases}$$

(i) Find the value of k.

(ii) Sketch the graph of $f(x)$.

(iii) According to this model, how many of the 100 cars would be expected to fall into each of the four categories?

(iv) Do you think the model fits the data well?

(v) Are there any obvious weaknesses in the model? If you were the planning officer, would you be prepared to accept the model as it is, or would you want any further information?

10 A fish farmer has a very large number of trout in a lake. Before deciding whether to net the lake and sell the fish, she collects a sample of 100 fish and weighs them. The results (in kg) are as follows.

Weight, W	Frequency	Weight, W	Frequency
$0 < W \leqslant 0.5$	2	$2.0 < W \leqslant 2.5$	27
$0.5 < W \leqslant 1.0$	10	$2.5 < W \leqslant 3.0$	12
$1.0 < W \leqslant 1.5$	23	$3.0 < W$	0
$1.5 < W \leqslant 2.0$	26		

(i) Illustrate these data on a histogram, with the number of fish on the vertical scale and W on the horizontal scale. Is the distribution of the data symmetrical, positively skewed or negatively skewed?

A friend of the farmer suggests that W can be modelled as a continuous random variable and proposes four possible probability density functions.

$$f_1(w) = \tfrac{2}{9} w(3 - w) \qquad f_2(w) = \tfrac{10}{81} w^2 (3 - w)^2$$

$$f_3(w) = \tfrac{4}{27} w^2 (3 - w) \qquad f_4(w) = \tfrac{4}{27} w(3 - w)^2$$

in each case for $0 \leqslant w \leqslant 3$.

(ii) Using your calculator (or otherwise), sketch the curves of the four p.d.f.s and state which one matches the data most closely in general shape.

(iii) Use this p.d.f. to calculate the number of fish which that model predicts should fall within each group.

(iv) Do you think it is a good model?

11 During a war the crew of an aeroplane has to destroy an enemy railway line by dropping bombs. The distance between the railway line and where the bomb hits the ground is X m, where X has the following p.d.f.

$$f(x) = \begin{cases} 10^{-4}(a + x) & \text{for } -a \leqslant x \leqslant 0 \\ 10^{-4}(a - x) & \text{for } 0 \leqslant x \leqslant a \\ 0 & \text{otherwise.} \end{cases}$$

(i) Find the value of a.

(ii) Find $P(50 \leqslant X \leqslant 60)$.

(iii) Find $P(|X| < 20)$.

[MEI]

12 A random variable X has a probability density function f given by

$$f(x) = \begin{cases} cx(5-x) & 0 \leqslant x \leqslant 5 \\ 0 & \text{otherwise.} \end{cases}$$

Show that $c = \frac{6}{125}$.

The lifetime X (in years) of an electric light bulb has this distribution. Given that a standard lamp is fitted with two such new bulbs and that their failures are independent, find the probability that neither bulb fails in the first year and the probability that exactly one bulb fails within two years.

[**MEI**]

13 This graph shows the probability distribution function, $f(x)$, for the heights, X, of waves at the point with Latitude 44°N Longitude 41°W.

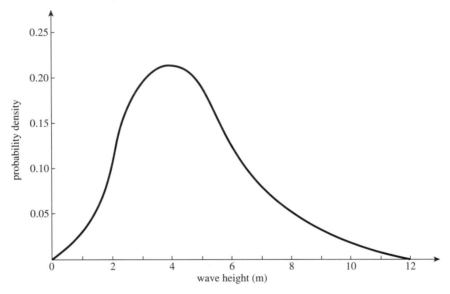

(i) Write down the values of $f(x)$ when $x = 0, 2, 4, \ldots, 12$.

(ii) Hence estimate the probability that the height of a randomly selected wave is in the interval

    (a) 0–2 m   **(b)** 2–4 m   **(c)** 4–6 m

    (d) 6–8 m   **(e)** 8–10 m   **(f)** 10–12 m.

A model is proposed in which

$$\begin{aligned} f(x) &= kx(12-x)^2 \quad \text{for } 0 \leqslant x \leqslant 12 \\ &= 0 \qquad\qquad\;\; \text{otherwise.} \end{aligned}$$

(iii) Find the value of k.

(iv) Find, according to this model, the probability that a randomly selected wave is in the interval

    (a) 0–2 m   **(b)** 2–4 m   **(c)** 4–6 m

    (d) 6–8 m   **(e)** 8–10 m   **(f)** 10–12 m.

(v) By comparing the figures from the model with the real data, state whether you think it is a good model or not.

Expectation and variance

You will recall that, for a discrete random variable, expectation and variance are given by:

$$E(X) = \sum_i x_i p_i$$

$$\text{Var}(X) = \sum_i (x_i - \mu)^2 p_i = \sum_i x_i^2 p_i - [E(X)]^2$$

where μ is the mean and p_i is the probability of the outcome x_i for $i = 1, 2, 3, \ldots,$ with the various outcomes covering all possibilities.

The expressions for the expectation and variance of a continuous random variable are equivalent, but with summation replaced by integration.

$$E(X) = \int_{\substack{\text{All} \\ \text{values} \\ \text{of } x}} x f(x) \, dx$$

$$\text{Var}(X) = \int_{\substack{\text{All} \\ \text{values} \\ \text{of } x}} (x - \mu)^2 f(x) \, dx = \int_{\substack{\text{All} \\ \text{values} \\ \text{of } x}} x^2 f(x) \, dx - [E(X)]^2$$

$E(X)$ is the same as the population mean, μ, and is often called the mean of X.

EXAMPLE 1.4

The response time, in seconds, for a contestant in a general knowledge quiz is modelled by a continuous random variable X whose p.d.f. is

$$f(x) = \frac{x}{50} \quad \text{for } 0 < x \leqslant 10.$$

The rules state that a contestant who makes no answer is disqualified from the whole competition. This has the consequence that everybody gives an answer, if only a guess, to every question. Find

(i) the mean time in seconds for a contestant to respond to a particular question
(ii) the standard deviation of the time taken.

The organiser estimates the proportion of contestants who are guessing by assuming that they are those whose time is at least one standard deviation greater than the mean.

(iii) Using this assumption, estimate the probability that a randomly selected response is a guess.

SOLUTION

(i) Mean time: $E(X) = \int_0^{10} x \dfrac{x}{50} \, dx$

$$= \left[\frac{x^3}{150} \right]_0^{10} = \frac{1000}{150} = \frac{20}{3}$$

$$= 6\tfrac{2}{3}$$

The mean time is $6\tfrac{2}{3}$ seconds.

(ii) Variance: $\text{Var}(X) = \int_0^{10} x^2 f(x) \, dx - [E(X)]^2$

$$= \int_0^{10} \frac{x^3}{50} \, dx - (6\tfrac{2}{3})^2$$

$$= \left[\frac{x^4}{200} \right]_0^{10} - (6\tfrac{2}{3})^2$$

$$= 5\tfrac{5}{9}$$

Standard deviation $= \sqrt{\text{Variance}} = \sqrt{5.\dot{5}}$

The standard deviation of the times is 2.357 seconds (to 3 dp).

(iii) All those with response times greater than $6.667 + 2.357 = 9.024$ seconds are taken to be guessing. The longest possible time is 10 seconds.

The probability that a randomly selected response is a guess is given by

$$\int_{9.024}^{10} \frac{x}{50} \, dx$$

$$= \left[\frac{x^2}{100} \right]_{9.024}^{10}$$

$$= 0.186$$

So just under 1 in 5 answers are deemed to be guesses.

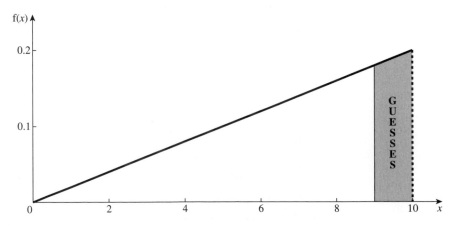

Figure 1.6

Note

Although the intermediate answers have been given rounded to three decimal places, more figures have been carried forward into subsequent calculations.

EXAMPLE 1.5

The number of hours per day that Darren spends in his garden is modelled (as on page 7) by the continuous random variable X, the p.d.f. of which is

$$f(x) = \frac{1}{9}x \qquad \text{for } 0 \leqslant x \leqslant 3$$

$$= \frac{(6-x)}{9} \qquad \text{for } 3 < x \leqslant 6$$

$$= 0 \qquad \text{otherwise.}$$

Find $E(X)$, the mean number of hours per day that Darren spends in his garden.

SOLUTION

$$E(X) = \int_{-\infty}^{\infty} x f(x)\, dx$$

$$= \int_0^3 x \frac{1}{9}x\, dx + \int_3^6 x \frac{(6-x)}{9}\, dx$$

$$= \left[\frac{x^3}{27}\right]_0^3 + \left[\frac{x^2}{3} - \frac{x^3}{27}\right]_3^6$$

$$= 1 + (12 - 8) - (3 - 1)$$

$$= 3$$

Darren spends a mean of 3 hours per day in his garden.

Notice that in this case $E(X)$ can be found from the line of symmetry of the graph of $f(x)$. This situation often arises and you should be alert to the possibility of finding $E(X)$ by symmetry; see figure 1.7.

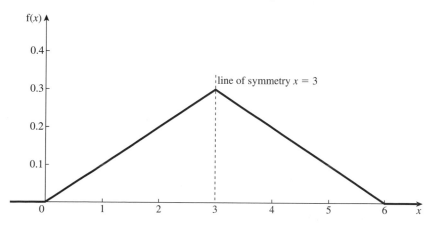

Figure 1.7

The median

The median value of a continuous random variable X with p.d.f. $f(x)$ is the value m for which

$$P(X < m) = P(X > m) = 0.5.$$

Consequently $\displaystyle\int_{-\infty}^{m} f(x)\,dx = 0.5$ and $\displaystyle\int_{m}^{\infty} f(x)\,dx = 0.5.$

The median is the value m such that the line $x = m$ divides the area under the curve $f(x)$ into two equal parts. In figure 1.8 a is the smallest possible value of X, b the largest. The line $x = m$ divides the shaded region into two regions A and B, both with area 0.5.

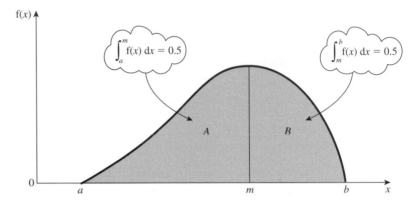

Figure 1.8

Note

In general the mean does not divide the area into two equal parts but it will do so if the curve is symmetrical about it because, in that case, it is equal to the median.

The mode

The mode of a continuous random variable X whose p.d.f. is $f(x)$ is the value for which $f(x)$ has the greatest value. Thus the mode is the value of X where the curve is at its highest.

If the mode is at a local maximum of $f(x)$, then it may often be found by differentiating $f(x)$ and solving the equation

$$f'(x) = 0.$$

For which of the distributions in figure 1.9 could the mode be found by differentiating the p.d.f.?

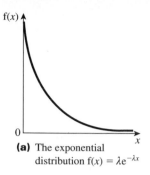

(a) The exponential
distribution $f(x) = \lambda e^{-\lambda x}$

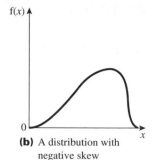

(b) A distribution with
negative skew

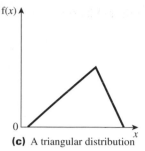

(c) A triangular distribution

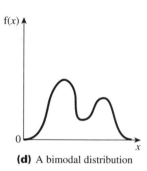

(d) A bimodal distribution

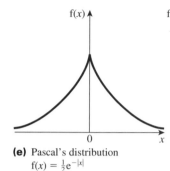

(e) Pascal's distribution
$f(x) = \frac{1}{2}e^{-|x|}$

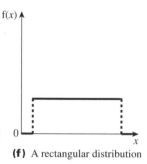

(f) A rectangular distribution

Figure 1.9

EXAMPLE 1.6

The continuous random variable X has p.d.f. $f(x)$ where

$$f(x) = 4x(1 - x^2) \quad \text{for } 0 \leqslant x \leqslant 1$$
$$= 0 \qquad\qquad \text{otherwise.}$$

Find **(i)** the mode

     (ii) the median.

SOLUTION

(i) The mode is found by differentiating $f(x) = 4x - 4x^3$

$$f'(x) = 4 - 12x^2$$

Solving $f'(x) = 0$ $x = \dfrac{1}{\sqrt{3}} = 0.577$ to 3 decimal places.

> $x = -0.577$ is also a root of $f'(x) = 0$ but is outside the range $0 \leqslant x \leqslant 1$

It is easy to see from the shape of the graph (see figure 1.10 overleaf) that this must be a maximum, and so the mode is 0.577.

(ii) The median, m, is given by $\displaystyle\int_{-\infty}^{m} f(x)\,dx = 0$

$$\Rightarrow \int_{0}^{m} (4x - 4x^3)\,dx = 0.5 \qquad \longleftarrow \boxed{\text{Since } m \geqslant 0}$$

$$\left[2x^2 - x^4\right]_{0}^{m} = 0.5$$

$$2m^2 - m^4 = 0.5$$

Rearranging gives

$$2m^4 - 4m^2 + 1 = 0.$$

This is a quadratic equation in m^2. The formula gives

$$m^2 = \frac{4 \pm \sqrt{16 - 8}}{4}$$

$$m = 0.541 \text{ or } 1.307 \text{ to 3 decimal places.}$$

Since 1.307 is outside the domain of X, the median is 0.541.

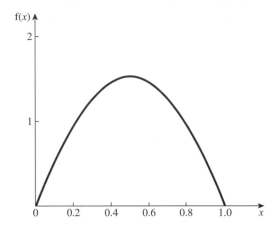

Figure 1.10

The rectangular distribution

It is common to describe distributions by the shapes of the graphs of their p.d.f.s: U-shaped, J-shaped, etc.

The *rectangular distribution* is particularly simple since its p.d.f. is constant over a range of values and zero elsewhere.

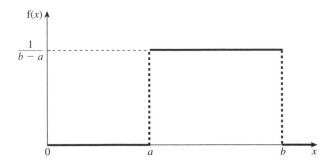

Figure 1.11

In figure 1.11, X may take values between a and b, and is zero elsewhere. Since the area under the graph must be 1, the height is $\dfrac{1}{b-a}$. The rectangular distribution is sometimes referred to as the uniform distribution, but this term is more often used when all the outcomes for a *discrete* variable are equally likely, like the score when a single die is thrown.

EXAMPLE 1.7

A junior gymnastics league is open to children who are at least five years old but have not yet had their ninth birthday. The age, X years, of a member is modelled by the rectangular distribution over the range of possible values between five and nine. Age is measured in years and decimal parts of a year, rather than just completed years. Find

(i)   the p.d.f. $f(x)$ of X
(ii) $P(6 \leqslant X \leqslant 7)$
(iii) $E(X)$
(iv) $Var(X)$
(v)   the percentage of the children whose ages are within one standard deviation of the mean.

SOLUTION

(i)   The p.d.f. $f(x) = \dfrac{1}{9-5} = \dfrac{1}{4}$ for $5 \leqslant x < 9$

$= 0$ otherwise.

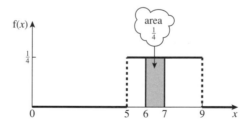

Figure 1.12

(ii) $P(6 \leqslant X \leqslant 7) = \frac{1}{4}$ by inspection of the rectangle above.
Alternatively, using integration

$$P(6 \leqslant X \leqslant 7) = \int_6^7 f(x)\,dx = \int_6^7 \frac{1}{4}\,dx$$

$$= \left[\frac{x}{4}\right]_6^7$$

$$= \frac{7}{6} - \frac{6}{4}$$

$$= \frac{1}{4}.$$

(iii) By the symmetry of the graph $E(X) = 7$. Alternatively, using integration

$$E(X) = \int_{-\infty}^{\infty} x f(x)\, dx = \int_5^9 \frac{x}{4}\, dx$$

$$= \left[\frac{x^2}{8}\right]_5^9$$

$$= \frac{81}{8} - \frac{25}{8} = 7.$$

(iv) $\mathrm{Var}(X) = \int_{-\infty}^{\infty} x^2 f(x)\, dx - [E(X)]^2 = \int_5^9 \frac{x^2}{4}\, dx - 7^2$

$$= \left[\frac{x^3}{12}\right]_5^9 - 49$$

$$= \frac{729}{12} - \frac{125}{12} - 49$$

$$= 1.333 \text{ to 3 decimal places.}$$

(v) Standard deviation $= \sqrt{\mathrm{Variance}} = \sqrt{1.333} = 1.155.$

So the percentage within 1 standard deviation of the mean is

$$\frac{2 \times 1.155}{4} \times 100\% = 57.7\%.$$

? What percentage would be within 1 standard deviation of the mean for a normal distribution? Why is the percentage less in this example?

The mean and variance of the rectangular distribution

In the previous example the mean and variance of a particular rectangular distribution were calculated. This can easily be extended to the general rectangular distribution given by:

$$f(x) = \frac{1}{b-a} \quad \text{for } a \leqslant x \leqslant b$$

$$= 0 \qquad \text{otherwise.}$$

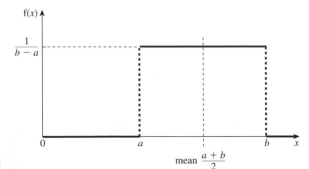

Figure 1.13

Mean   By symmetry the mean is $\dfrac{a+b}{2}$.

Variance $\text{Var}(X) = \displaystyle\int_{-\infty}^{\infty} x^2 f(x)\,dx - [E(X)]^2$

$$= \int_{a}^{b} x^2 f(x)\,dx - [E(X)]^2$$

$$= \int_{a}^{b} \frac{x^2}{b-a}\,dx - \left(\frac{a+b}{2}\right)^2$$

$$= \left[\frac{x^3}{3(b-a)}\right]_{a}^{b} - \frac{1}{4}(a^2 + 2ab + b^2)$$

$$= \frac{b^3 - a^3}{3(b-a)} - \frac{1}{4}(a^2 + 2ab + b^2)$$

$$= \frac{(b-a)}{3(b-a)}(b^2 + ab + a^2) - \frac{1}{4}(a^2 + 2ab + b^2)$$

$$= \frac{1}{12}(b^2 - 2ab + a^2)$$

$$= \frac{1}{12}(b-a)^2$$

The exponential distribution

This distribution is often used to model the waiting times between events, such as earthquakes, radioactive emissions, telephone calls, etc., when the events being counted occur at random, singly and independently, at a constant overall rate.

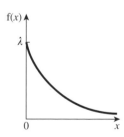

Figure 1.14

The random variable X has p.d.f. $f(x)$ as shown in figure 1.14, given by:

$$f(x) = \lambda e^{-\lambda x} \quad \text{for } x \geqslant 0$$
$$= 0 \qquad \text{otherwise.}$$

The mean is $\dfrac{1}{\lambda}$ and the variance is $\dfrac{1}{\lambda^2}$.

PROOF

(i) Show that, for $x \geqslant 0$, $f(x)$ satisfies the requirements for a p.d.f., that is $f(x) \geqslant 0$ for all x and

$$\int_{\substack{\text{All} \\ \text{values} \\ \text{of } x}} f(x)\,dx = 1.$$

The graph above shows that $f(x) \geqslant 0$ for all x.

$$\int_{\substack{\text{All} \\ \text{values} \\ \text{of } x}} f(x)\,dx = \int_0^\infty \lambda e^{-\lambda x}\,dx$$

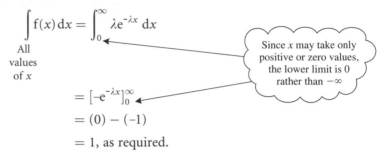

Since x may take only positive or zero values, the lower limit is 0 rather than $-\infty$

$$= [-e^{-\lambda x}]_0^\infty$$

$$= (0) - (-1)$$

$$= 1, \text{ as required.}$$

(ii) Show that $E(X) = \dfrac{1}{\lambda}$.

$$E(X) = \int_0^\infty x f(x)\,dx = \int_0^\infty \lambda x e^{-\lambda x}\,dx$$

Integrating by parts gives

$$E(X) = [-x e^{-\lambda x}]_0^\infty + \int_0^\infty e^{-\lambda x}\,dx$$

$$= 0 - \left[\frac{e^{-\lambda x}}{\lambda}\right]_0^\infty$$

$$= -\left(-\frac{1}{\lambda}\right)$$

$$= \frac{1}{\lambda}, \text{ as required.}$$

(iii) Show that $\mathrm{Var}(X) = \dfrac{1}{\lambda^2}$.

$$\mathrm{Var}(X) = \int_0^\infty x^2 f(x)\,dx - [E(X)]^2 = \int_0^\infty x^2 \lambda e^{-\lambda x}\,dx - \left(\frac{1}{\lambda}\right)^2$$

Integrating by parts gives

$$\mathrm{Var}(X) = [-x^2 e^{-\lambda x}]_0^\infty + \int_0^\infty e^{-\lambda x} 2x\,dx - \frac{1}{\lambda^2}$$

$$= 0 + 2\int_0^\infty x e^{-\lambda x}\,dx - \frac{1}{\lambda^2}.$$

Since it has been shown in part (ii) that

$$\int_0^\infty \lambda x e^{-\lambda x}\, dx = \frac{1}{\lambda}$$

it follows that

$$2\int_0^\infty x e^{-\lambda x}\, dx = \frac{2}{\lambda^2}$$

and so

$$\mathrm{Var}(X) = \frac{2}{\lambda^2} - \frac{1}{\lambda^2}$$

$$= \frac{1}{\lambda^2}, \text{ as required.}$$

EXERCISE 1B

1 The continuous random variable X has p.d.f. $f(x)$ where

$$f(x) = \tfrac{1}{8}x \quad \text{for } 0 \leqslant x \leqslant 4$$
$$= 0 \qquad \text{otherwise.}$$

Find

(i) $\mathrm{E}(X)$

(ii) $\mathrm{Var}(X)$

(iii) the median value of X.

2 The continuous random variable T has p.d.f. defined by

$$f(t) = \frac{6-t}{18} \quad \text{for } 0 \leqslant t \leqslant 6$$
$$= 0 \qquad \text{otherwise.}$$

Find

(i) $\mathrm{E}(T)$

(ii) $\mathrm{Var}(T)$

(iii) the median value of T.

3 The continuous random variable Y has p.d.f. $f(y)$ defined by

$$f(y) = 12y^2(1-y) \quad \text{for } 0 \leqslant y \leqslant 1$$
$$= 0 \qquad \text{otherwise.}$$

Find

(i) $\mathrm{E}(Y)$

(ii) $\mathrm{Var}(Y)$

(iii) the median value of Y.

4 The random variable X has p.d.f.

$$f(x) = \tfrac{1}{6} \quad \text{for } -2 \leqslant x \leqslant 4$$
$$= 0 \quad \text{otherwise.}$$

(i) Sketch the graph of $f(x)$.

(ii) Find $\mathrm{P}(X < 2)$.

(iii) Find $\mathrm{E}(X)$.

(iv) Find $\mathrm{P}(|X| < 1)$.

5 The continuous random variable X has p.d.f. $f(x)$ defined by

$$f(x) = \begin{cases} \frac{2}{9}x(3-x) & \text{for } 0 \leqslant x \leqslant 3 \\ 0 & \text{otherwise.} \end{cases}$$

  (i) Find $E(X)$.

  (ii) Find $Var(X)$.

  (iii) Find the mode of X.

  (iv) Find the median value of X.

  (v) Draw a sketch graph of $f(x)$ and comment on your answers to parts (i), (iii) and (iv) in the light of what it shows you.

6 The random variable X has a rectangular distribution over the interval $(-2, 5)$. Find

  (i) the p.d.f. of X

  (ii) $E(X)$

  (iii) $Var(X)$

  (iv) $P(X$ is positive$)$.

7 The function $f(x) = \begin{cases} k(3+x) & \text{for } 0 \leqslant x \leqslant 2 \\ 0 & \text{otherwise.} \end{cases}$

  is the probability density function of the random variable X.

  (i) Show that $k = \frac{1}{8}$.

  (ii) Find the mean and variance of X.

  (iii) Find the probability that a randomly selected value of X lies between 1 and 2.

8 A continuous random variable X has a rectangular distribution over the interval $(4, 7)$. Find

  (i) the p.d.f. of X

  (ii) $E(X)$

  (iii) $Var(X)$

  (iv) $P(4.1 \leqslant X \leqslant 4.8)$.

9 The distribution of the lengths of adult Martian lizards is uniform between 10 cm and 20 cm. There are no adult lizards outside this range.

  (i) Write down the p.d.f. of the lengths of the lizards.

  (ii) Find the mean and variance of the lengths of the lizards.

  (iii) What proportion of the lizards have lengths within

    (a) one standard deviation of the mean

    (b) two standard deviations of the mean?

10 The p.d.f. of the lifetime, X hours, of a brand of electric light bulb is modelled by

$$f(x) = \begin{cases} \dfrac{1}{60\,000}x & \text{for } 0 \leqslant x \leqslant 300 \\ \dfrac{1}{50} - \dfrac{1}{20\,000}x & \text{for } 300 < x \leqslant 400 \\ 0 & \text{for } x > 400. \end{cases}$$

(i) Sketch the graph of f(x).

(ii) Show that f(x) fulfils the conditions for it to be a p.d.f.

(iii) Find the expected lifetime of a bulb.

(iv) Find the variance of the lifetimes of the bulbs.

(v) Find the probability that a randomly selected bulb will last less than 100 hours.

11 The marks of candidates in an examination are modelled by the continuous random variable X with p.d.f.

$$f(x) = kx(x - 50)^2(100 - x) \quad \text{for } 0 \leqslant x \leqslant 100$$
$$= 0 \qquad\qquad\qquad \text{otherwise.}$$

(i) Find the value of k.

(ii) Sketch the graph of f(x).

(iii) Describe the shape of the graph and give an explanation of how such a graph might occur, in terms of the examination and the candidates.

(iv) Is it permissible to model a mark, which is a discrete variable going up in steps of 1, by a continuous random variable like X, as defined in this question?

12 The municipal tourism officer at a Mediterranean resort on the Costa Del Sol wishes to model the amount of sunshine per day during the holiday season. She denotes by X the number of hours of sunshine per day between 8 am and 8 pm and she suggests the following probability density function for X:

$$f(x) = k[(x - 3)^2 + 4] \quad \text{for } 0 \leqslant x \leqslant 12$$
$$= 0 \qquad\qquad\qquad \text{otherwise.}$$

(i) Show that $k = \frac{1}{300}$ and sketch the graph of the p.d.f. f(x).

(ii) Assuming that the model is accurate, find the mean and standard deviation of the number of hours of sunshine per day. Find also the probability of there being more than eight hours of sunshine on a randomly chosen day.

(iii) Obtain a cubic equation for m, the median number of hours of sunshine, and verify that m is about 9.74 to 2 decimal places.

[MEI]

13 The continuous random variable X has p.d.f. f(x) defined by

$$f(x) = ae^{-kx} \quad \text{for } x \geqslant 0$$
$$= 0 \qquad\quad \text{otherwise.}$$

Find, in terms of k,

(i) a

(ii) $E(X)$

(iii) $Var(X)$

(iv) the median value of X

(v) There are many situations where this random variable might be used as a model. Describe one such situation.

14 A statistician is also a keen cyclist. He believes that the distance which he cycles between punctures may be modelled by the random variable, X km, with p.d.f. $f(x)$ given by

$$f(x) = 0.005e^{-0.005x} \quad \text{for } x > 0$$
$$= 0 \quad \text{otherwise.}$$

(i) Find the mean distance he cycles between punctures.

(ii) He has just repaired one puncture. What is the probability that he will travel at least 500 km before having another one?

(iii) He has just repaired one puncture. What is the probability that he will travel less than 30 km before having another one?

On one occasion he starts a race with new tyres but then has a puncture after 30 km. When he starts again he has another puncture after k km. He says that according to his model the combined probability of first a puncture within 30 km and then one within k km is 0.005.

(iv) What is the value of k?

15 The continuous random variable X has p.d.f. $f(x)$ defined by

$$f(x) = \begin{cases} \dfrac{a}{x} & \text{for } 1 \leqslant x \leqslant 2 \\ 0 & \text{otherwise.} \end{cases}$$

(i) Find the value of a.

(ii) Sketch the graph of $f(x)$.

(iii) Find the mean and variance of X.

(iv) Find the proportion of values of X between 1.5 and 2.

(v) Find the median value of X.

16 An examination is taken by a large number of candidates. The marks scored are modelled by the continuous random variable X with probability density function

$$f(x) = kx^3(120 - x), \quad 0 \leqslant x \leqslant 100.$$

(You should assume throughout this question that marks are on a continuous scale. Hence there is no need to consider continuity corrections.)

(i) Sketch the graph of this probability density function. What does the model suggest about the abilities of the candidates in relation to this examination?

(ii) Show that $k = 10^{-9}$.

(iii) The pass mark is set at 50. Find what proportion of candidates fail the examination.

(iv) The top 20% of candidates are awarded a distinction. Determine whether a mark of 90 is sufficient for a distinction. Find the least whole number mark which is sufficient for a distinction.

[MEI]

Expectation and variance of a function of X

There are times when one random variable is a function of another random variable. For example:

- as part of an experiment you are measuring temperatures in Celsius but then need to convert them to Fahrenheit: $F = 1.8C + 32$;
- you are measuring the lengths of the sides of square pieces of material and deducing their areas: $A = L^2$;
- you are estimating the ages, A years, of hedgerows by counting the number, n, of types of shrubs and trees in 30 m lengths: $A = 100n - 50$.

In fact in any situation where you are entering the value of a random variable into a formula, the outcome will be another random variable which is a function of the one you entered. Under these circumstances you may need to find the expectation and variance of such a function of a random variable.

For a discrete random variable, X, the expectation of a function $g[X]$ is given by:

$$E(g[X]) = \Sigma g[x_i] p(x_i),$$

$$Var(g[X]) = \Sigma (g[x_i])^2 p(x_i) - \{E(g[X])\}^2.$$

The equivalent results for a continuous random variable, X, with p.d.f. $f(x)$ are:

$$E(g[X]) = \int_{\substack{\text{All} \\ \text{values} \\ \text{of } x}} g[x]\, f(x)\, dx,$$

$$Var(g[X]) = \int_{\substack{\text{All} \\ \text{values} \\ \text{of } x}} (g[x])^2 f(x)\, dx - \{E(g[X])\}^2.$$

You may find it helpful to think of the function $g[X]$ as a new variable, say Y.

EXAMPLE 1.8   The continuous random variable X has p.d.f. $f(x)$ given by:

$$f(x) = \frac{x}{50} \quad \text{for } 0 \leqslant x \leqslant 10$$

$$= 0 \quad \text{otherwise.}$$

(This random variable was used to model response times in Example 1.4.)

(i)   Find $E(3X + 4)$.
(ii) Find $3E(X) + 4$.
(iii) Find $Var(3X + 4)$.
(iv) Verify that $Var(3X + 4) = 3^2 Var(X)$.

SOLUTION

(i) $E(3X + 4) = \int_0^{10} (3x + 4) \dfrac{x}{50} \, dx$

$= \int_0^{10} \dfrac{1}{50} (3x^2 + 4x) \, dx$

$= \left[\dfrac{x^3}{50} + \dfrac{x^2}{25} \right]_0^{10}$

$= 20 + 4$

$= 24$

(ii) $3E(X) + 4 = 3 \int_0^{10} x \dfrac{x}{50} \, dx + 4$

$= \left[\dfrac{3}{150} x^3 \right]_0^{10} + 4$

$= 20 + 4$

$= 24$

Notice here that $E(3X + 4) = 24 = 3E(X) + 4$.

(iii) To find $\text{Var}(3X + 4)$, use

$$\text{Var}[g(X)] = \int [g(x)]^2 f(x) \, dx - \{E[g(X)]\}^2$$

with $g(X) = 3X + 4.$

$\text{Var}(3X + 4) = \int_0^{10} (3x + 4)^2 \dfrac{1}{50} x \, dx - 24^2$

$= \int_0^{10} \dfrac{1}{50} (9x^3 + 24x^2 + 16x) \, dx - 576$

$= \dfrac{1}{50} \left[\dfrac{9x^4}{4} + 8x^3 + 8x^2 \right]_0^{10} - 576$

$= 50$

(iv) $\text{Var}(X) = E(X^2) - [E(X)]^2$

$E(X^2) = \int_0^{10} x^2 \dfrac{1}{50} x \, dx$ $\qquad$ $E(X) = \int_0^{10} x \dfrac{1}{50} x \, dx$

$E(X^2) = \int_0^{10} \dfrac{1}{50} x^3 \, dx$ $\qquad$ $E(X) = \int_0^{10} \dfrac{1}{50} x^2 \, dx$

$E(X^2) = \left[\dfrac{1}{200} x^4 \right]_0^{10}$ $\qquad$ $E(X) = \left[\dfrac{1}{150} x^3 \right]_0^{10}$

$E(X^2) = 50$ $\qquad\qquad$ $E(X) = 6.\dot{6}$

$\text{Var}(X) = 50 - 6.\dot{6}^2 = 5.\dot{5}$

$3^2 \text{Var}(X) = 9 \times 5.\dot{5} = 50$

From part (iii), $\text{Var}(3X + 4) = 50$

So $\text{Var}(3X + 4) = 3^2 \text{Var}(X)$ as required.

General results

This example illustrates a number of general results for random variables, continuous or discrete.

$$E(c) = c \qquad\qquad \mathrm{Var}(c) = 0$$

$$E(aX) = aE(X) \qquad\qquad \mathrm{Var}(aX) = a^2\mathrm{Var}(X)$$

$$E(aX + b) = aE(X) + b \qquad \mathrm{Var}(aX + b) = a^2\mathrm{Var}(X)$$

$$E[g(X) + h(X)] = E[g(X)] + E[h(X)]$$

where a, b and c are constants and $g(X)$ and $h(X)$ are functions of X.

Note

The result $\mathrm{Var}(aX) = a^2\mathrm{Var}(X)$ should not surprise you. It follows from the fact that if the standard deviation of a set of data is k then the standard deviation of the set formed by multiplying all the data by a constant, a, is ak. That is,

$$\text{standard deviation}(aX) = a \times \text{standard deviation}(X)$$

Since variance $=$ (standard deviation)^2

then $\quad \mathrm{Var}(aX) = [\text{standard deviation}(aX)]^2$

$$= [a \times \text{standard deviation}(X)]^2$$

$$= a^2[\text{standard deviation}(X)]^2$$

$$= a^2\,\mathrm{Var}(X).$$

EXAMPLE 1.9

The continuous random variable X has p.d.f. $f(x)$ given by

$$f(x) = \frac{3}{125}x^2 \quad \text{for } 0 \leqslant x \leqslant 5$$

$$= 0 \qquad\quad \text{otherwise.}$$

Find **(i)** $E(X)$ **(ii)** $\mathrm{Var}(X)$ **(iii)** $E(7X - 3)$ **(iv)** $\mathrm{Var}(7X - 3)$.

SOLUTION

(i) $\displaystyle E(X) = \int_{\infty}^{\infty} x f(x)\,dx$

$$= \int_{0}^{5} x \frac{3}{125}x^2\,dx$$

$$= \left[\frac{3}{500}x^4\right]_{0}^{5}$$

$$= 3.75$$

(ii) $\text{Var}(X) = \int_{-\infty}^{\infty} x^2 f(x)\,dx - [E(X)]^2$

$\qquad = \int_{0}^{5} x^2 \dfrac{3}{125} x^2\,dx - 3.75^2$

$\qquad = \left[\dfrac{3}{625} x^5\right]_{0}^{5} - 14.0625$

$\qquad = 15 - 14.0625$

$\qquad = 0.9375$

(iii) $E(7X - 3) = 7E(X) - 3$

$\qquad\qquad = 7 \times 3.75 - 3$

$\qquad\qquad = 23.25$

(iv) $\text{Var}(7X - 3) = 7^2 \text{Var}(X)$

$\qquad\qquad = 49 \times 0.9375$

$\qquad\qquad = 45.94$ to 2 decimal places.

1 The number of kilograms of metal extracted from 10 kg of ore from a certain mine is modelled by a continuous random variable X with probability density function $f(x)$, where $f(x) = cx(2 - x)^2$ if $0 \leqslant x \leqslant 2$ and $f(x) = 0$ otherwise, where c is a constant.

Show that c is $\frac{3}{4}$, and find the mean and variance of X.

The cost of extracting the metal from 10 kg of ore is £10x. Find the expected cost of extracting the metal from 10 kg of ore.

2 A continuous random variable X has the p.d.f.:

$$f(x) = k \quad \text{for } 0 \leqslant x \leqslant 5$$
$$\quad = 0 \quad \text{otherwise.}$$

(i) Find the value of k.

(ii) Sketch the graph of $f(x)$.

(iii) Find $E(X)$.

(iv) Find $E(4X - 3)$ and show that your answer is the same as $4E(X) - 3$.

3 The continuous random variable X has p.d.f.

$$f(x) = 4x^3 \quad \text{for } 0 \leqslant x \leqslant 1$$
$$\quad = 0 \quad \text{otherwise.}$$

(i) Find $E(X)$.

(ii) Find $E(X^2)$.

(iii) Find $\text{Var}(X)$.

(iv) Verify that $E(5X + 1) = 5E(X) + 1$.

4 A continuous random variable Y has p.d.f.

$$f(y) = \tfrac{2}{9} y(3 - y) \quad \text{for } 0 \leqslant y \leqslant 3$$
$$= 0 \qquad\qquad \text{otherwise.}$$

  (i) Find $E(Y)$.

  (ii) Find $E(Y^2)$.

  (iii) Find $E(Y^2) - [E(Y)]^2$.

  (iv) Find $E(2Y^2 + 3Y + 4)$.

  (v) Find $\int_0^3 (y - E(Y))^2 \, f(y) \, dy$. Why is the answer the same as that for part (iii)?

5 A continuous random variable X has p.d.f. $f(x)$, where

$$f(x) = 12x^2(1 - x) \quad \text{for } 0 \leqslant x \leqslant 1$$
$$= 0 \qquad\qquad \text{otherwise.}$$

  (i) Find μ, the mean of X.

  (ii) Find $E(6X - 7)$ and show that your answer is the same as $6E(X) - 7$.

  (iii) Find the standard deviation of X.

  (iv) What is the probability that a randomly selected value of X lies within one standard deviation of μ?

6 The continuous random variable X has p.d.f.

$$f(x) = \tfrac{2}{25}(7 - x) \quad \text{for } 2 \leqslant x \leqslant 7$$
$$= 0 \qquad\qquad \text{otherwise.}$$

  The function $g(X)$ is defined by $g(x) = 3x^2 + 4x + 7$.

  (i) Find $E(X)$.

  (ii) Find $E[g(X)]$.

  (iii) Find $E(X^2)$ and hence find $3E(X^2) + 4E(X) + 7$.

  (iv) Use your answers to parts (ii) and (iii) to verify that

$$E[g(X)] = 3E(X^2) + 4E(X) + 7.$$

7 A toy company sells packets of coloured plastic equilateral triangles. The triangles are actually offcuts from the manufacture of a totally different toy, and the length, X, of one side of a triangle may be modelled as a random variable with a rectangular distribution for $2 \leqslant x \leqslant 8$.

  (i) Find the p.d.f. of X.

  (ii) An equilateral triangle of side x has area a. Find the relationship between a and x.

  (iii) Find the probability that a randomly selected triangle has area greater than $15 \, \text{cm}^2$.

  (iv) Find the expectation and variance of the area of a triangle.

8 A continuous random variable has probability density function f defined by

$$f(x) = kx(4 - x) \quad \text{for } 0 \leqslant x \leqslant 4$$
$$= 0 \qquad\qquad \text{otherwise.}$$

  Evaluate k and the mean of the distribution.

A particle moves along a straight line in such a way that during the first four seconds of its motion its velocity at time t seconds is $v\,\text{ms}^{-1}$, where

$$v = 2(t + 1).$$

The particle is observed at time t seconds, where t denotes a random value from a distribution whose probability density function is the function f defined above. Calculate the probability that at the time of observation the velocity of the particle is less than $4\,\text{ms}^{-1}$. **[JMB]**

9 A canon fires balls at a fixed angle of elevation with an initial speed U in ms^{-1}. The horizontal distance they travel R (in m), is given by

$$R = 0.1\,U^2.$$

The initial speed of the canon balls depends on the charge and so is somewhat variable.

(i) The initial speed U is modelled as having a rectangular distribution between 80 and 120 in ms^{-1}.

   (a) Find the mean and standard deviation of U.
   (b) Find the mean and standard deviation of R.

(ii) An alternative model is considered for U and this is shown in the diagram below.

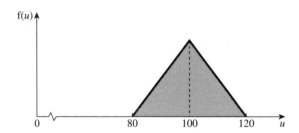

   (a) Find the p.d.f. for U under this model.
   (b) Find the mean and variance of U.
   (c) Find the mean and variance of R.

10 A company specialises in extinguishing oil fires. For much of the time its personnel have no work to do but they have to be available to travel anywhere at a moment's notice. The company reckons its overhead costs are £5000 per day, whether they have work or not.

The company model the waiting time, t days, between jobs by the exponential distribution

$$f(t) = \frac{1}{100}\,e^{-\left(\frac{1}{100}\right)t}.$$

   (i) Find the mean time that they wait between jobs.
   (ii) Find the probability that they go for one year or more between jobs.
   (iii) Find the number of days, x, within which they have a 95% probability of getting another job.

When they work, they take an average of 20 days to put out a fire. They aim to make enough profit during that time to be able to meet their costs for the next x days, so that there is a probability of 0.95 that they will have another job before their profit from the last one has run out.

(iv) How much should they charge per day when on a job?

(v) What is the expectation of their profit per day?

11 Every day I travel to and from work on the local shuttle bus, which runs every 10 minutes. The time I have to wait for the bus is modelled by the random variable T, which has a rectangular distribution on the interval $[0, 10]$.

(i) Write down the probability density function for T, and state its mean and variance.

The *total* time I have to wait for a bus, going to and coming from work, is modelled by the random variable X whose probability density function is given by

$$
\begin{aligned}
f(x) &= 0.01x & \text{for } 0 \leqslant x \leqslant 10 \\
&= 0.01(20 - x) & \text{for } 10 < x \leqslant 20 \\
&= 0 & \text{otherwise.}
\end{aligned}
$$

(ii) Sketch the graph of the probability density function for X.

(iii) State $E(X)$ and use integration to find $\text{Var}(X)$.

The times I wait for the bus when going to work and coming home from work are represented by independent random variables T_1 and T_2 respectively, so that $X = T_1 + T_2$.

(iv) Find $P(X \geqslant 14)$. Give a reason why you would expect $P(X \geqslant 14)$ to be greater than $P(T_1 \geqslant 7) \times P(T_2 \geqslant 7)$.

[**MEI**]

12 The continuous random variable X has probability density function

$$f(x) = \tfrac{3}{1024}x(x - 8)^2 \quad 0 \leqslant x \leqslant 8.$$

A sketch of $f(x)$ is shown in the diagram.

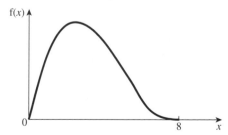

(i) Find $E(X)$ and show that $\text{Var}(X) = 2.56$.

The times, in minutes, taken by a doctor to see her patients are modelled by the continuous random variable $T = X + 2$.

(ii) Sketch the distribution of T and describe in words what this model implies about the lengths of the doctor's appointments.

[**MEI**]

Cumulative distribution function

An entrant for the competition proposed a model in which a runner's time, X hours, is a continuous random variable with p.d.f.

$$f(x) = \tfrac{4}{27}(x-1)(4-x)^2 \quad 1 \leqslant x \leqslant 4$$
$$= 0 \qquad\qquad\qquad \text{otherwise}$$

According to this model the mode is at 2 hours, and everybody finishes in between 1 hour and 4 hours; see figure 1.15.

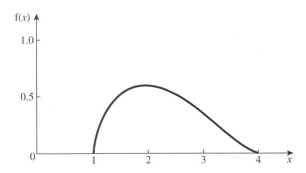

Figure 1.15

How does this model compare with the figures you were given for the actual race?

Those figures gave the *cumulative distribution*, the total numbers (expressed as percentages) of runners who had finished by certain times. To obtain the equivalent figures from the model, you must find the relevant area under the graph in figure 1.16.

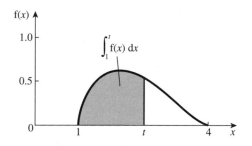

Figure 1.16

In this model, the proportion finishing by time t hours is given by

$$\int_1^t f(x)\,dx = \int_1^t \tfrac{4}{27}(x-1)(4-x)^2\,dx$$

$$= \tfrac{4}{27}\int_1^t (x^3 - 9x^2 + 24x - 16)\,dx$$

$$= \tfrac{4}{27}\left[\tfrac{1}{4}x^4 - 3x^3 + 12x^2 - 16x\right]_1^t$$

$$= \tfrac{4}{27}\left(\tfrac{1}{4}t^4 - 3t^3 + 12t^2 - 16t\right) - (-1)$$

$$= \tfrac{1}{27}t^4 - \tfrac{4}{9}t^3 + \tfrac{16}{9}t^2 - \tfrac{64}{27}t + 1$$

This is called the *cumulative distribution function* and denoted by $F(t)$.
In this case,

$$F(t) = 0 \qquad\qquad\qquad\qquad\quad \text{for } t < 1$$

$$= \tfrac{1}{27}t^4 - \tfrac{4}{9}t^3 + \tfrac{16}{9}t^2 - \tfrac{64}{27}t + 1 \quad \text{for } 1 \leqslant t \leqslant 4$$

$$= 1 \qquad\qquad\qquad\qquad\qquad \text{for } t > 0.$$

To find the proportions of runners finishing by any time, substitute that value for t; so when $t = 2$

$$F(2) = \tfrac{1}{27} \times 2^4 - \tfrac{4}{9} \times 2^3 + \tfrac{16}{9} \times 2^2 - \tfrac{64}{27} \times 2 + 1$$

$$= 0.41 \text{ to 2 decimal places.}$$

Here is the complete table, with all the values worked out.

Time (hours)	Model	Runners
1.00	0.00	0.00
1.25	0.04	0.03
1.50	0.13	0.15
1.75	0.26	0.33
2.00	0.41	0.49
2.25	0.55	0.57
2.50	0.69	0.75
3.00	0.89	0.91
3.50	0.98	0.99
4.00	1.00	1.00

Notice the distinctive shape of the curves of these functions (figure 1.17), sometimes called an *ogive*. You have probably met this already, when drawing cumulative frequency curves, for example in *Statistics 1*.

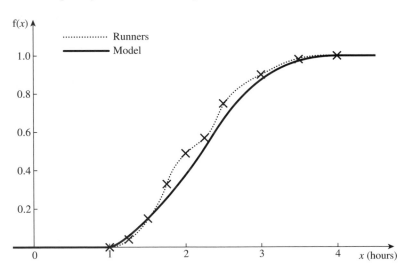

Figure 1.17

❓ Do you think that this model is worth the £150 prize? If you were on the organising committee what more might you look for in a model?

Notes

1 Notice the use of lower and upper case letters here. The probability density function is denoted by the lower case f, whereas the cumulative distribution function is given the upper case F.

2 F was derived here as a function of t rather than x, to avoid using the same variable in the expression to be integrated. In this case t was a natural variable to use because time was involved, but that is not always the case.

It is more usual to write F as a function of x, F(x), but you would not be correct to write down an expression like

$$F(x) = \int_1^x \frac{4}{27}(x-1)(4-x)^2 \, dx \qquad \text{INCORRECT}$$

since x would then be both a limit of the integral and the variable used within it.

To overcome this problem a dummy variable, u, is used in the rest of this section, so that F(x) is now written,

$$F(x) = \int_1^x \frac{4}{27}(u-1)(4-u)^2 \, du \qquad \text{CORRECT}$$

You may of course use another symbol, like y or p, rather than u, anything except x.

3 The term cumulative distribution function is often abbreviated to c.d.f.

Properties of the cumulative distribution function, F(x)

The graphs, figure 1.18, show the probability density function f(x) and the cumulative distribution function F(x) of a typical continuous random variable X. You will see that the values of the random variable always lie between a and b.

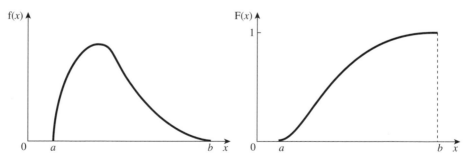

Figure 1.18

These graphs illustrate a number of general results for cumulative distribution functions.

1 F(x) = 0   for $x \leqslant a$, the lower limit of x.

The probability of X taking a value less than or equal to a is zero; the value of X must be greater than or equal to a.

2 $F(x) = 1$ for $x \geqslant b$, the upper limit of x.
X cannot take values greater than b.

3 $P(c \leqslant X \leqslant d) = F(d) - F(c)$
$P(c \leqslant X \leqslant d) = P(X \leqslant d) - P(X \leqslant c)$

This is very useful when finding probabilities from a p.d.f. or a c.d.f.

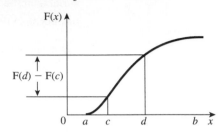

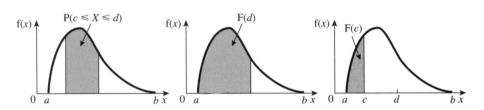

Figure 1.19

4 The median, m, satisfies the equation $F(m) = 0.5$.
$P(X \leqslant m) = 0.5$ by definition of the median.

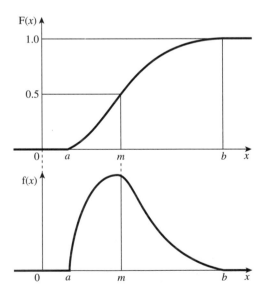

Figure 1.20

5 $f(x) = \dfrac{\mathrm{d}}{\mathrm{d}x} F(x) = F'(x)$

Since you integrate $f(x)$ to obtain $F(x)$, the reverse must also be true:
differentiating $F(x)$ gives $f(x)$.

EXAMPLE 1.10

A machine saws planks of wood to a nominal length. The continuous random variable X represents the error in millimetres of the actual length of a plank coming off the machine. The variable X has p.d.f. $f(x)$ where

$$f(x) = \frac{10 - x}{50} \quad \text{for } 0 \leqslant x \leqslant 10$$

$$= 0 \qquad \text{otherwise.}$$

(i) Sketch $f(x)$.

(ii) Find the cumulative distribution function $F(x)$.

(iii) Sketch $F(x)$ for $0 \leqslant x \leqslant 10$.

(iv) Find $P(2 \leqslant X \leqslant 7)$.

(v) Find the median value of X.

A customer refuses to accept planks for which the error is greater than 8 mm.

(vi) What percentage of planks will he reject?

SOLUTION

(i)

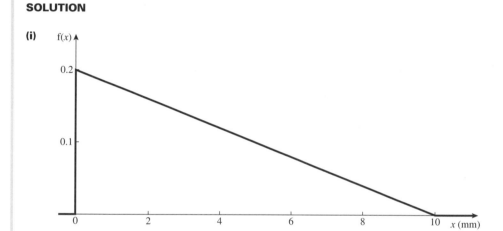

Figure 1.21

(ii) $\displaystyle F(x) = \int_0^x \frac{(10 - u)}{50}\, du$

$$= \frac{1}{50} \left[10u - \frac{u^2}{2} \right]_0^x$$

$$= \frac{1}{5}x - \frac{1}{100}x^2$$

The full definition of $F(x)$ is:

$$F(x) = 0 \qquad\qquad \text{for } x < 0$$

$$= \frac{1}{5}x - \frac{1}{100}x^2 \quad \text{for } 0 \leqslant x \leqslant 10$$

$$= 1 \qquad\qquad \text{for } x > 10.$$

(iii) The graph F(x) is shown in figure 1.22.

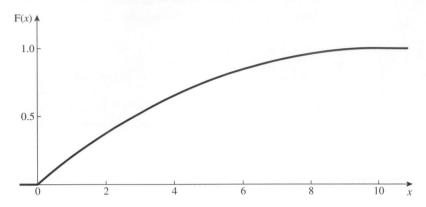

Figure 1.22

(iv) $P(2 \leqslant X \leqslant 7) = F(7) - F(2)$

$= \left[\frac{7}{5} - \frac{49}{100}\right] - \left[\frac{2}{5} - \frac{4}{100}\right]$

$= 0.91 - 0.36$

$= 0.55.$

(v) The median value of X is found by solving the equation

$$F(m) = 0.5$$

$$\tfrac{1}{5} m - \tfrac{1}{100} m^2 = 0.5.$$

This is rearranged to give

$$m^2 - 20m + 50 = 0$$

$$m = \frac{20 \pm \sqrt{20^2 - 4 \times 50}}{2}$$

$$m = 2.93 \text{ (or 17.07, outside the domain for } X\text{).}$$

The median error is 2.93 mm.

(vi) The customer rejects those planks for which $8 \leqslant X \leqslant 10$

$$P(8 \leqslant X \leqslant 10) = F(10) - F(8)$$

$$= 1 - 0.96$$

so 4% of planks are rejected.

EXAMPLE 1.11

The p.d.f. of a continuous random variable X is given by:

$$f(x) = \frac{x}{24} \qquad \text{for } 0 \leqslant x \leqslant 4$$

$$= \frac{(12 - x)}{48} \qquad \text{for } 4 \leqslant x \leqslant 12$$

$$= 0 \qquad \text{otherwise.}$$

(i) Sketch $f(x)$.

(ii) Find the cumulative distribution function $F(x)$.

(iii) Sketch $F(x)$.

SOLUTION

(i) The graph of $f(x)$ is shown in figure 1.23.

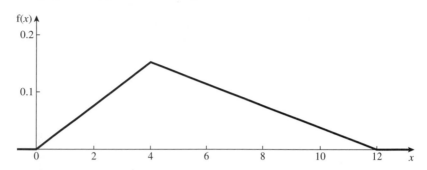

Figure 1.23

(ii) For $0 \leqslant x \leqslant 4$, $F(x) = \displaystyle\int_0^x \frac{u}{24} \, du$

$$= \left[\frac{u^2}{48} \right]_0^x$$

$$= \frac{x^2}{48}$$

and so $F(4) = \frac{1}{3}$.

For $4 \leqslant x \leqslant 12$, a second integration is required:

$$F(x) = \int_0^4 \frac{u}{24} \, du + \int_4^x \left(\frac{12 - u}{48} \right) du$$

$$= F(4) + \left[\frac{u}{4} - \frac{u^2}{96} \right]_4^x$$

$$= \frac{1}{3} + \frac{x}{4} - \frac{x^2}{96} - \frac{5}{6}$$

$$= -\frac{1}{2} + \frac{x}{4} - \frac{x^2}{96}$$

So the full definition of $F(x)$ is

$$
\begin{aligned}
F(x) &= 0 & &\text{for } x < 0 \\
&= \frac{x^2}{48} & &\text{for } 0 \leqslant x \leqslant 4 \\
&= -\frac{1}{2} + \frac{x}{4} - \frac{x^2}{96} & &\text{for } 4 \leqslant x \leqslant 12 \\
&= 1 & &\text{for } x > 12.
\end{aligned}
$$

(iii) The graph of F(x) is shown in figure 1.24.

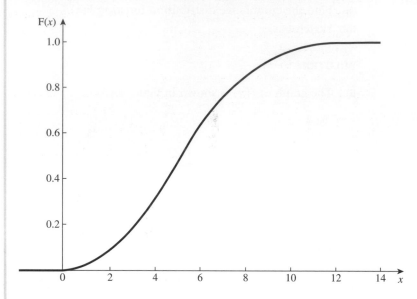

Figure 1.24

EXAMPLE 1.12

The continuous random variable X has cumulative distribution function $F(x)$ given by:

$$F(x) = \begin{cases} 0 & \text{for } x < 2 \\ \dfrac{x^2}{32} - \dfrac{1}{8} & \text{for } 2 \leqslant x \leqslant 6 \\ 1 & \text{for } x > 6. \end{cases}$$

Find the p.d.f. $f(x)$.

SOLUTION

$$f(x) = \frac{d}{dx} F(x)$$

$$f(x) = \begin{cases} \dfrac{d}{dx} F(x) = 0 & \text{for } x < 2 \\ \dfrac{d}{dx} F(x) = \dfrac{x}{16} & \text{for } 2 \leqslant x \leqslant 6 \\ \dfrac{d}{dx} F(x) = 0 & \text{for } x > 6. \end{cases}$$

Finding the p.d.f. of a function of a continuous random variable

The cumulative distribution function provides you with a stepping stone between the p.d.f. of a continuous random variable and that of a function of that variable. Example 1.13 shows how it is done.

EXAMPLE 1.13

1

Cumulative distribution function

A company make metal boxes to order. The basic process consists of cutting four squares off the corners of a sheet of metal, which is then folded and welded along the joins. Consequently, for every box there are four square offcuts of waste metal.

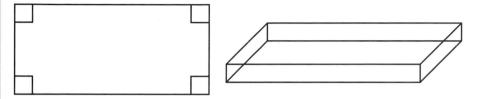

Figure 1.25

The company is looking for ways to cut costs and the designers wonder if anything can be done with these square pieces. They decide in the first place to investigate the distribution of their sizes. A survey of the large pile in their scrap area shows that they vary in length up to a maximum of 2 decimetres. It is suggested their lengths can be modelled as a continuous random variable L with probability density function

$$f(l) = \frac{l}{4}(4 - l^2) \quad \text{for } 0 < l \leqslant 2$$
$$= 0 \qquad\qquad \text{otherwise.}$$

Assume this model to be accurate.

(i) Find the cumulative distribution function for the length of a square.

(ii) Hence derive the cumulative distribution function for the area of a square.

(iii) Find the p.d.f. for the area of a square.

(iv) Sketch the graphs of the probability density functions and the cumulative distribution functions of the length and the area.

(v) Find the mean area of the square offcuts when making a box.

SOLUTION

(i) The c.d.f. is

$$F(l) = \int_0^l \frac{u}{4}(4 - u^2)\,du$$

$$= \left[\frac{u^2}{2} - \frac{u^4}{16}\right]_0^l$$

> Notice the use of u as a dummy variable for L

and so

$$F(l) = 0 \qquad\qquad \text{for } l < 0$$
$$= \frac{l^2}{2} - \frac{l^4}{16} \quad \text{for } 0 < l \leqslant 2$$
$$= 1 \qquad\qquad \text{for } l > 2.$$

(ii) The area, a, and length, l, of a square are related by

$$a = l^2$$

and since $0 < l \leqslant 2$

it follows that $0 < a \leqslant 2^2$, that is, $0 < a \leqslant 4$.

Substituting a for l^2 in the answer to part (i), and using the appropriate range of values for a, gives the cumulative distribution function $H(a)$, because

$$H(a) = P(A \leqslant a)$$
$$= P(L^2 \leqslant l^2)$$
$$= P(L \leqslant l)$$
$$= F(l)$$

Since $\qquad F(l) = \dfrac{l^2}{2} - \dfrac{l^4}{16} \quad$ for $0 < l \leqslant 2$

it follows that $\quad H(a) = \dfrac{a}{2} - \dfrac{a^2}{16} \quad$ for $0 < a \leqslant 4$

$$= 1 \qquad\qquad \text{for } a > 4$$

Note

Notice the use of H and h for the c.d.f. and the p.d.f. of the area, in place of F and f. The different letters are used to distinguish these from the corresponding functions for the length.

(iii) The p.d.f. for the area of a square is found by differentiating $H(a)$

$$h(a) = \frac{\mathrm{d}}{\mathrm{d}a} H(a) = \frac{1}{2} - \frac{a}{8} \quad \text{for } 0 < a \leqslant 4$$
$$= 0 \qquad\qquad \text{otherwise.}$$

(iv) The graphs of the p.d.f.s of the length and the area are shown in figure 1.26.

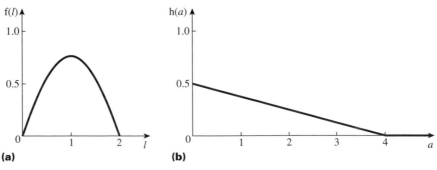

(a)

(b)

Figure 1.26

(a) $f(l) = \dfrac{1}{4} l(4 - l^2) \quad$ for $0 < l \leqslant 2$

$\qquad = 0 \qquad\qquad$ otherwise

(b) $h(a) = \dfrac{1}{2} - \dfrac{a}{8} \quad$ for $0 < a \leqslant 4$

$\qquad = 0 \qquad\qquad$ otherwise

The graphs of the c.d.f.s of the length and the area are shown in figure 1.27.

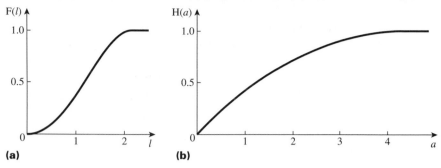

Figure 1.27

(a) $F(l) = \dfrac{l^2}{2} - \dfrac{l^4}{16}$ for $0 < l \leqslant 2$

$\quad\quad\quad = 1$ for $l > 2$.

(b) $H(a) = \dfrac{a}{2} - \dfrac{a^2}{16}$ for $0 < a \leqslant 4$

$\quad\quad\quad = 1$ for $a > 4$.

(v) Mean $= E(A) = \displaystyle\int_0^4 a h(a)\, da$

$= \displaystyle\int_0^4 \left(\dfrac{a}{2} - \dfrac{a^2}{8} \right) da$

$= \left[\dfrac{a^2}{4} - \dfrac{a^3}{24} \right]_0^4$

$= \dfrac{4}{3}$

Note

This could also have been found as the mean of a function of a continuous random variable, using the general result

$$E[g(X)] = \int_{\substack{\text{All} \\ \text{values} \\ \text{of } x}} g(x)\, f(x)\, dx$$

where x is the length (not the area) of one of the squares.

In this case $g(x) = x^2$ $\quad$ $f(x) = \dfrac{x}{4}(4 - x^2)$ and $0 < x \leqslant 2$,

giving $\quad E[g(X)] = \displaystyle\int_0^2 x^2 \dfrac{x}{4}(4 - x^2)\, dx$

$= \displaystyle\int_0^2 \left(x^3 - \dfrac{x^5}{4} \right) dx$

$= \left[\dfrac{x^4}{4} - \dfrac{x^6}{24} \right]_0^2$

$= 4 - \dfrac{64}{24} = \dfrac{4}{3}$

i.e. the same answer.

The normal distribution

The continuous random variable with which you are probably most familiar is Z, which has the standardised normal distribution.

Remember that a value z is calculated from the actual data value, x, by using the transformation

$$z = \frac{x - \mu}{\sigma}$$

where μ is the population mean and σ is the population standard deviation.

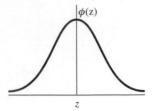

Figure 1.28

The p.d.f. of the normal curve is given the special notation $\phi(z)$ and is given by

$$\phi(z) = \frac{1}{\sqrt{2\pi}} \, e^{-\frac{1}{2}z^2} \quad \text{for } -\infty < z < \infty.$$

Consequently the cumulative distribution function, which is given the notation $\Phi(z)$, is given by

$$\Phi(z) = \frac{1}{\sqrt{2\pi}} \int_{-\infty}^{z} e^{-\frac{1}{2}u^2} \, du$$

The function $\Phi(z)$ represents the probability of a value of Z less than or equal to z. That is

$$\Phi(z) = P(Z \leqslant z)$$

Unfortunately the integration cannot be carried out algebraically and so there is no neat expression for $\Phi(z)$. Instead the integration has been performed numerically for different values of z, and the results are given in the form of the well-known normal distribution tables, see figure 1.29.

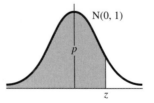

z	.00	.01	.02	.03	.04	.05	.06	.07	.08	.09	1	2	3	4	5	6	7	8	9
0.0	.5000	5040	5080	5120	5160	5199	5239	5279	5319	5359	4	8	12	16	20	24	28	32	36
0.1	.5398	5438	5478	5517	5557	5596	5636	5675	5714	5753	4	8	12	16	20	24	28	32	35
0.2	.5793	5832	5871	5910	5948	5987	6026	6064	6103	6141	4	8	12	15	19	23	27	31	35
0.3	.6179	6217	6255	6293	6331	6368	6406	6443	6480	6517	4	8	11	15	19	23	26	30	34
0.4	.6554	6951	6628	6664	6700	6736	6772	6808	6844	6879	4	7	11	14	18	22	25	29	32
0.5	.6915	6950	6985	7019	7054	7088	7123	7157	7190	7224	3	7	10	14	17	21	24	27	31
0.6	.7257	7291	7324	7357	7389	7422	7454	7486	7517	7549	3	6	10	13	16	19	23	26	29
0.7	.7580	761·					7764	7794	7823	7852	3	6	9	12	15	18	21	24	27
0.8	7··						78	8106	8133		3	6	°	··	··	·7	19	22	25

Figure 1.29

The table gives the probability p of a random variable distributed as $N(0, 1)$ being less than z.

1 The continuous random variable X has p.d.f. $f(x)$ where

$$f(x) = 0.2 \quad \text{for } 0 \leqslant x \leqslant 5$$
$$= 0 \quad \text{otherwise.}$$

(i) Find $E(X)$.

(ii) Find the cumulative distribution function, $F(x)$.

(iii) Find $P(0 \leqslant x \leqslant 2)$ using **(a)** $F(x)$ **(b)** $f(x)$ and show your answer is the same by each method.

2 The continuous random variable U has p.d.f. $f(u)$ where

$$f(u) = ku \quad \text{for } 5 \leqslant u \leqslant 8$$
$$= 0 \quad \text{otherwise.}$$

(i) Find the value of k.

(ii) Sketch $f(u)$.

(iii) Find $F(u)$.

(iv) Sketch the graph of $F(u)$.

3 A continuous random variable X has p.d.f. $f(x)$ where

$$f(x) = cx^2 \quad \text{for } 1 \leqslant x \leqslant 4$$
$$= 0 \quad \text{otherwise.}$$

(i) Find the value of c.

(ii) Find $F(x)$.

(iii) Find the median of X.

(iv) Find the mode of X.

4 The continuous random variable X has p.d.f. $f(x)$ given by

$$f(x) = \frac{k}{(x+1)^4} \quad \text{for } x \geqslant 0$$
$$= 0 \quad \text{for } x < 0$$

where k is a constant.

(i) Show that $k = 3$, and find the cumulative distribution function.

(ii) Find also the value of x such that $P(X < x) = \frac{7}{8}$.

[Cambridge]

5 The continuous random variable X has c.d.f. given by

$$F(x) = \begin{cases} 0 & \text{for } x < 0 \\ 2x - x^2 & \text{for } 0 \leqslant x \leqslant 1 \\ 1 & \text{for } x > 1. \end{cases}$$

(i) Find $P(X > 0.5)$.

(ii) Find the value of q such that $P(X < q) = \frac{1}{4}$.

(iii) Find the p.d.f. $f(x)$ of X, and sketch its graph.

[Cambridge]

6 The continuous random variable X has p.d.f. $f(x)$ given by

$$f(x) = \begin{cases} k(4 - x^2) & \text{for } 0 \leqslant x \leqslant 2 \\ 0 & \text{otherwise} \end{cases}$$

where k is a constant.

Show that $k = \frac{3}{16}$ and find the values of $E(X)$ and $\text{Var}(X)$.

Find the cumulative distribution function for X, and verify by calculation that the median value of X is between 0.69 and 0.70.

[Cambridge]

7 A random variable X has p.d.f. $f(x)$ where

$$f(x) = 12x^2(1 - x) \quad \text{for } 0 \leqslant x \leqslant 1$$

and $\qquad\qquad f(x) = 0 \qquad\qquad\qquad$ for all other x.

Find μ, the mean of X, and show that σ, the standard deviation of X, is $\frac{1}{5}$. Show that $F(x)$, the probability that $X \leqslant x$ (for any value of x between 0 and 1), satisfies

$$F(x) = \begin{cases} 0 & \text{for } x < 0 \\ 4x^3 - 3x^4 & \text{for } 0 \leqslant x \leqslant 1 \\ 1 & \text{for } x > 1. \end{cases}$$

Use this result to show that $P(|X - \mu| < \sigma) = 0.64$.

What would this probability be if, instead, X were normally distributed?

[MEI]

8 The temperature in degrees Celsius in a refrigerator which is operating properly has probability density function given by

$$f(t) = \begin{cases} kt^2(12 - t) & 0 < t < 12 \\ 0 & \text{otherwise.} \end{cases}$$

(i) Show that the value of k is $\frac{1}{1728}$.
(ii) Find the cumulative distribution function $F(t)$.
(iii) Show, by substitution, that the median temperature is about 7.37 °C.
(iv) The temperature in a refrigerator is too high if it is over 10 °C. Find the probability that this occurs.

[MEI]

9 The probability that a randomly chosen flight from Stanston Airport is delayed by more than x hours is

$$\frac{(x - 10)^2}{100} \quad \text{for } 0 \leqslant x \leqslant 10.$$

No flights leave early, and none is delayed for more than 10 hours. The delay, in hours, for a randomly chosen flight is denoted by X.

(i) Find the median, m, of X, correct to three significant figures.
(ii) Find the cumulative distribution function, F, of X and sketch the graph of F.
(iii) Find the probability density function, f, of X, and sketch the graph of f.
(iv) Show that $E(X) = \frac{10}{3}$.

[Cambridge]

10 A random variable X has p.d.f. $f(x)$, where

$$f(x) = k\sin2x \quad \text{for } 0 \leqslant x \leqslant \frac{\pi}{2}$$
$$= 0 \qquad \text{otherwise.}$$

By integration find, in terms of x and the constant k, an expression for the cumulative distribution function of X for $0 \leqslant x \leqslant \frac{\pi}{2}$.

Hence show that $k = 1$ and find the probability that $X < \frac{\pi}{8}$.

[MEI]

11 On any day, the amount of time, measured in hours, that Mr Goggle spends watching television is a continuous random variable T, with cumulative distribution function given by

$$F(t) = \begin{cases} 0 & t < 0 \\ 1 - k(15 - t)^2 & 0 \leqslant t \leqslant 15 \\ 1 & t > 15 \end{cases}$$

where k is a constant.

(i) Show that $k = \frac{1}{225}$ and find $P(5 \leqslant T \leqslant 10)$.

(ii) Show that, for $0 \leqslant t \leqslant 15$, the probability density function of T is given by

$$f(t) = \frac{2}{15} - \frac{2t}{225}.$$

(iii) Find the median of T.

(iv) Find $\text{Var}(T)$.

[Cambridge]

12 The time, T minutes, between customer arrivals at a country store, from Monday to Friday, can be modelled, for $t \geqslant 0$, by the probability density function $f(t)$ defined by

$$f(t) = 0.1\,e^{-0.1t}.$$

(i) Find the probability that the time between arrivals is
  (a) less than 5 minutes   **(b)** more than 15 minutes
  illustrating these probabilities on a sketch of the graph of $f(t)$.

(ii) Obtain the cumulative distribution function for T. Hence find the median time between arrivals.

The time, T minutes, between customer arrivals at the country store on Saturdays can be modelled by the probability density function $g(t)$ defined by

$$g(t) = \lambda e^{-\lambda t}$$

where λ is a positive constant.

(iii) On Saturdays, the probability of the time between customer arrivals exceeding 5 minutes is 0.4. Estimate the value of λ.

[MEI]

13 A firm has a large number of employees. The distance in miles they have to travel each day from home to work can be modelled by a continuous random variable X whose **cumulative distribution function** is given by

$$F(1) = 0$$

$$F(x) = k\left(1 - \frac{1}{x}\right) \quad 1 \leqslant x \leqslant b$$

$$F(b) = 1$$

where b represents the farthest distance anybody lives from work.

The diagram below shows a sketch of this cumulative distribution function.

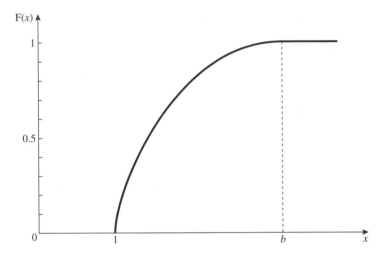

A survey suggests that $b = 5$. Use this parameter for parts (i) to (iv).

(i) Show that $k = 1.25$.

(ii) Write down and solve an equation to find the median distance travelled to work.

(iii) Find the probability that an employee lives within half a mile of the median.

(iv) Derive the probability density function for X and illustrate it with a sketch.

(v) Show that, for any value of b greater than 1, the median distance travelled does not exceed 2.

[**MEI**]

14 A random variable X is distributed uniformly between 0 and a. Write down the expectation of X and determine its standard deviation by integration.

Two independent observations of X are made, X_1 and X_2. By taking X_1 and X_2 as cartesian co-ordinates and considering the lines $X_1 + X_2 = Y$, show that the probability that

$(X_1 + X_2) \leqslant Y$ is $\dfrac{Y^2}{2a^2}$ if $0 \leqslant Y \leqslant a$ and is $1 - \dfrac{(2a - Y)^2}{2a^2}$ if $a \leqslant Y \leqslant 2a$.

Hence obtain the probability density function of $(X_1 + X_2)$. [**MEI**]

15 A continuous random variable X has probability density function $f(x)$. The probability that $X \leqslant x$ is given by the function $F(x)$.

Explain why $F'(x) = f(x)$.

A rod of length $2a$ is broken into two parts at a point whose position is random. State the form of the probability distribution of the length of the smaller part, and state also the mean value of this length.

Two equal rods, each of length $2a$, are broken into two parts at points whose positions are random. X is the length of the shortest of the four parts thus obtained. Find the probability, $F(x)$, that $X \leqslant x$, where $0 < x \leqslant a$.

Hence, or otherwise, show that the probability density function of X is given by

$$f(x) = \frac{2(a - x)}{a^2} \quad \text{for } 0 < x \leqslant a$$
$$= 0 \quad \text{for } x \leqslant 0, \, x > a.$$

Show that the mean value of X is $\frac{1}{3}a$.

Write down the mean value of the sum of the two smaller parts and show that the mean values of the four parts are in the proportions $1:2:4:5$.

[JMB]

16 (i) Explain the significance of the results:

(a) $\displaystyle\int_{-\infty}^{\infty} \frac{1}{\sqrt{2\pi}} \, e^{-\frac{1}{2}z^2} \, dz = 1$ **(b)** $\displaystyle\int_{-\infty}^{\infty} \frac{1}{\sqrt{2\pi}} \, z e^{-\frac{1}{2}z^2} \, dz = 0.$

The random variable Y is given by $Y = Z^2$ (where Z is the standardised normal variable).

(ii) Using the results in part (i), find

(a) $E(Y)$ **(b)** Var (Y).

(You will need to use integration by parts.)

17 The random variable Y is given by $Y = Z^2$ (where Z is the standardised normal variable).

(i) State the range of values Y may take.

(ii) Explain carefully why $P(Y \leqslant y) = 2P(0 \leqslant Z \leqslant \sqrt{y})$.

Let $G(y)$ denote the cumulative distribution function for Y.

(iii) Show that $G(y) = 2(\Phi(\sqrt{y}) - \frac{1}{2})$.

(iv) Differentiate the result in part (iii) to show that the p.d.f. of Y is

$$g(y) = \frac{1}{\sqrt{2\pi y}} e^{-\frac{1}{2}y} \quad y \geqslant 0.$$

Note: for interest, the random variable Y has the χ^2 distribution with $\nu = 1$.

Continuous random variables

1

1 If X is a continuous random variable with p.d.f. $f(x)$

- $\int f(x)\,dx = 1$

- $f(x) \geqslant 0 \quad$ for all x

- $P(c \leqslant x \leqslant d = \int_c^d f(x)\,dx$

- $E(X) = \int xf(x)\,dx$

- $Var(X) = \int x^2 f(x)\,dx - [E(X)]^2$

- The mode of X is the value for which $f(x)$ has its greatest magnitude.

2 If $g[X]$ is a function of X then

- $E(g[X]) = \int g[x]f(x)\,dx$

- $Var(g[X]) = \int (g[x])^2 f(x)\,dx - [E(g[X])]^2$

3 **The cumulative distribution function**

- $F(x) = \int_a^x f(u)\,du$ where the constant a is the lower limit of X.

- $f(x) = \dfrac{d}{dx}F(x)$

- For the median, m, $F(m) = 0.5$

4 **The rectangular distribution over the interval (a, b)**

- $f(x) = \dfrac{1}{b-a}$

- $E(X) = \frac{1}{2}(a+b)$

- $Var(X) = \dfrac{(b-a)^2}{12}$

5 **The exponential distribution**

- $f(x) = \lambda e^{-\lambda x} \quad$ for $x \geqslant 0$

- $E(X) = \dfrac{1}{\lambda}$

- $Var(X) = \dfrac{1}{\lambda^2}$

6 **The normal distribution**

- $f(z) = \phi(z) = \dfrac{1}{\sqrt{2\pi}} e^{-\frac{1}{2}z^2} \quad -\infty < z < \infty$

- $F(z) = \Phi(z) = \displaystyle\int_{-\infty}^z \dfrac{1}{\sqrt{2\pi}} e^{-\frac{1}{2}u^2}\,du \quad -\infty < z < \infty$

2

Sums and differences of random variables

To approach zero defects, you must have statistical control of processes.

David Wilson

THE AVONFORD STAR

Unfair dismissal

'It was just one of those days', Janice Baptiste told the court. 'Everything went wrong. First the school bus arrived 5 minutes late to pick up my little boy. Then it was wet and slippery and there were so many people about that I just couldn't walk at my normal speed; usually I take 15 minutes but that day it took me 18 to get to work. And then when I got to work I had to wait $3\frac{1}{2}$ minutes for the lift instead of the usual $\frac{1}{2}$ minute. So instead of arriving my normal 10 minutes early I was 1 minute late.'

'Mrs Dickens just wouldn't listen', Janice went on. 'She said she did not employ people to make excuses and told me to leave there and then.'

Mrs Baptiste's bad morning turned a lot worse when her boss fired her

Like Janice, we all have days when everything goes wrong at once. There were three random variables involved in her arrival time at work: the time she had to wait for the school bus, S; the time she took to walk to work, W, and the time she had to wait for the lift, L.

Her total time for getting to work, T, was the sum of all three:

$$T = S + W + L.$$

Janice's case was essentially that the probability of T taking such a large value was very small. To estimate that probability you would need information about the distributions of the three random variables involved. You would also need to know how to handle the sum of two or more (in this case three) random variables, and that is the subject of this chapter.

EXAMPLE 2.1

The possible lengths (in cm) of the blades of cricket bats form a discrete uniform distribution:

38, 40, 42, 44, 46.

The possible lengths (in cm) of the handles of cricket bats also form a discrete uniform distribution:

22, 24, 26.

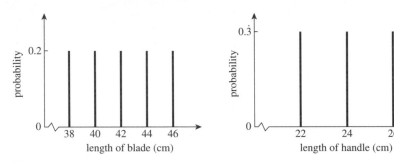

Figure 2.1

The blades and handles can be joined together to make bats of various lengths, and it may be assumed that the lengths of the two sections are independent.

(i) How many different (total) bat lengths are possible?

(ii) Work out the mean and variance of random variable X_1, the length (in cm) of the blades.

(iii) Work out the mean and variance of random variable X_2, the length (in cm) of the handles.

(iv) Work out the mean and variance of random variable $X_1 + X_2$, the total length of the bats.

(v) Verify that

$$E(X_1 + X_2) = E(X_1) + E(X_2)$$

and $$\text{Var}(X_1 + X_2) = \text{Var}(X_1) + \text{Var}(X_2).$$

SOLUTION

(i) The number of different bat lengths is 7. This can be seen from the sample space diagram below:

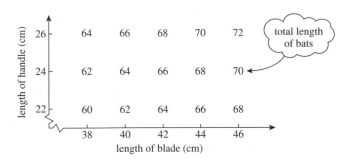

Figure 2.2

(ii)

Length of blade (cm)	38	40	42	44	46
Probability	0.2	0.2	0.2	0.2	0.2

$$E(X_1), \mu_1 = \Sigma xp = (38 \times 0.2) + (40 \times 0.2) + (42 \times 0.2) + (44 \times 0.2)$$
$$+ (46 \times 0.2)$$
$$= 42 \,\text{cm}$$

$$\text{Var}(X_1) = E(X_1^2) - \mu_1^2$$

$$E(X_1^2) = (38^2 \times 0.2) + (40^2 \times 0.2) + (42^2 \times 0.2) + (44^2 \times 0.2)$$

$$+ (46^2 \times 0.2)$$

$$= 1772$$

$$\text{Var}(X_1) = 1772 - 42^2 = 8$$

(iii)

Length of handle (cm)	22	24	26
Probability	$\frac{1}{3}$	$\frac{1}{3}$	$\frac{1}{3}$

$$E(X_2), \mu_2 = (22 \times \tfrac{1}{3}) + (24 \times \tfrac{1}{3}) + (26 \times \tfrac{1}{3}) = 24 \,\text{cm}$$

$$\text{Var}(X_2) = E(X_2^2) - \mu_2^2$$

$$E(X_2^2) = (22^2 \times \tfrac{1}{3}) + (24^2 \times \tfrac{1}{3}) + (26^2 \times \tfrac{1}{3}) = 578.667 \text{ to 3 dp}$$

$$\text{Var}(X_2) = 578.667 - 24^2 = 2.667 \text{ to 3 dp}$$

(iv) The probability distribution of $X_1 + X_2$ can be obtained from figure 2.2:

Total length of cricket bat (cm)	60	62	64	66	68	70	72
Probability	$\frac{1}{15}$	$\frac{2}{15}$	$\frac{3}{15}$	$\frac{3}{15}$	$\frac{3}{15}$	$\frac{2}{15}$	$\frac{1}{15}$

$$E(X_1 + X_2) = (60 \times \tfrac{1}{15}) + (62 \times \tfrac{2}{15}) + (64 \times \tfrac{3}{15}) + (66 \times \tfrac{3}{15}) + (68 \times \tfrac{3}{15})$$
$$+ (70 \times \tfrac{2}{15}) + (72 \times \tfrac{1}{15})$$
$$= 66 \,\text{cm}$$

$$\text{Var}(X_1 + X_2) = E[(X_1 + X_2)^2] - 66^2$$

$$E[(X_1 + X_2)^2] = (60^2 \times \tfrac{1}{15}) + (62^2 \times \tfrac{2}{15}) + (64^2 \times \tfrac{3}{15}) + (66^2 \times \tfrac{3}{15})$$

$$+ (68^2 \times \tfrac{3}{15}) + (70^2 \times \tfrac{2}{15}) + (72^2 \times \tfrac{1}{15})$$

$$= \frac{65\,500}{15} = 4366.667 \text{ to 3 dp}$$

$$\text{Var}(X_1 + X_2) = 4366.667 - 66^2 = 10.667 \text{ to 3 dp}$$

(v) $E(X_1 + X_2) = 66 = 42 + 24 = E(X_1) + E(X_2)$, as required.

$\text{Var}(X_1 + X_2) = 10.667 = 8 + 2.667 = \text{Var}(X_1) + \text{Var}(X_2)$, as required.

Note

You should notice that the standard deviations of X_1 and X_2 do not add up to the standard deviation of $X_1 + X_2$

$$\sqrt{8} + \sqrt{2.667} \neq \sqrt{10.667}$$

i.e. $\qquad 2.828 + 1.633 \neq 3.266$

General results

Example 2.1 has illustrated the following general results for the sums and differences of random variables.

For any two random variables X_1 and X_2

- $E(X_1 + X_2) = E(X_1) + E(X_2)$

Replacing X_2 by $-X_2$ in this result gives

$\quad E(X_1 + (-X_2)) = E(X_1) + E(-X_2)$

- $E(X_1 - X_2) = E(X_1) - E(X_2)$

If the variables X_1 and X_2 are independent then

- $\text{Var}(X_1 + X_2) = \text{Var}(X_1) + \text{Var}(X_2)$

Replacing X_2 by $-X_2$ gives

$\quad \text{Var}(X_1 + (-X_2)) = \text{Var}(X_1) + \text{Var}(-X_2)$

$\quad \text{Var}(X_1 + (-X_2)) = \text{Var}(X_1) + (-1)^2\text{Var}(X_2)$

- $\text{Var}(X_1 - X_2) = \text{Var}(X_1) + \text{Var}(X_2)$

The sums and differences of normal variables

If the variables X_1 and X_2 are normally distributed, then the distributions of $(X_1 + X_2)$ and $(X_1 - X_2)$ are also normal. The means of these distributions are $E(X_1) + E(X_2)$ and $E(X_1) - E(X_2)$.

You must, however, be careful when you come to their variances, since you may only use the result that

$$\text{Var}(X_1 \pm X_2) = \text{Var}(X_1) + \text{Var}(X_2)$$

to find the variances of these distributions if the variables X_1 and X_2 are independent.

This is the situation in the next two examples.

EXAMPLE 2.2

Robert Fisher, a keen chess player, visits his local club most days. The total time taken to drive to the club and back is modelled by a normal variable with mean 25 minutes and standard deviation 3 minutes. The time spent at the chess club is also modelled by a normal variable with mean 120 minutes and standard deviation 10 minutes. Find the probability that on a certain evening Mr Fisher is away from home for more than $2\frac{1}{2}$ hours.

SOLUTION

Let the random variable $X_1 \sim N(25, 3^2)$ represent the driving time, and the random variable $X_2 \sim N(120, 10^2)$ represent the time spent at the chess club.

Then the random variable T, where $T = X_1 + X_2 \sim N(145, (\sqrt{109})^2)$, represents his total time away.

So the probability that Mr Fisher is away for more than $2\frac{1}{2}$ hours (150 minutes) is given by

$$P(T > 150) = 1 - \Phi\left(\frac{150 - 145}{\sqrt{109}}\right)$$

$$= 1 - \Phi(0.479)$$

$$= 0.316$$

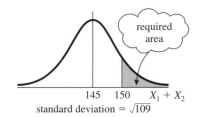

Figure 2.3

EXAMPLE 2.3

In the manufacture of a bridge made entirely from wood, circular pegs have to fit into circular holes. The diameters of the pegs are normally distributed with mean 1.60 cm and standard deviation 0.01 cm, while the diameters of the holes are normally distributed with mean 1.65 cm and standard deviation of 0.02 cm. What is the probability that a randomly chosen peg will not fit into a randomly chosen hole?

SOLUTION

Let the random variable X be the diameter of a hole:

$$X \sim N(1.65, 0.02^2) = N(1.65, 0.0004)$$

Let the random variable Y be the diameter of a peg:

$$Y \sim N(1.60, 0.01^2) = N(1.6, 0.0001)$$

Let $F = X - Y$. F represents the gap remaining between the peg and the hole and so the sign of F determines whether or not a peg will fit in a hole.

$$E(F) = E(X) - E(Y) = 1.65 - 1.60 = 0.05$$

$$Var(F) = Var(X) + Var(Y) = 0.0004 + 0.0001 = 0.0005$$

$$F \sim N(0.05, 0.0005)$$

If for any combination of peg and hole the value of F is negative, then the peg will not fit into the hole.

The probability that $F < 0$ is given by

$$\Phi\left(\frac{0 - 0.05}{\sqrt{0.0005}}\right) = \Phi(-2.236)$$

$$= 1 - 0.9873$$

$$= 0.0127$$

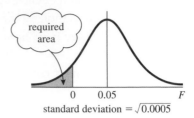

Figure 2.4

EXERCISE 2A

1 The menu at a cafe is shown below.

Main course		Dessert	
Fish and Chips	£3	Ice Cream	£1
Bacon and Eggs	£3.50	Apple Pie	£1.50
Pizza	£4	Sponge Pudding	£2
Steak and Chips	£5.50		

The owner of the cafe says that all the main-course dishes sell equally well, as do all the desserts, and that customers' choice of dessert is not influenced by the main course they have just eaten.

The variable M denotes the cost of the items for the main course, in pounds, and the variable D the cost of the items for the dessert. The variable T denotes the total cost of a two-course meal: $T = M + D$.

(i) Find the mean and variance of M.

(ii) Find the mean and variance of D.

(iii) List all the possible two-course meals, giving the price for each one.

(iv) Use your answer to part (iii) to find the mean and variance of T.

(v) Hence verify that for these figures

$$\text{mean } (T) = \text{mean } (M) + \text{mean } (D)$$

and   variance $(T) = $ variance $(M) + $ variance (D).

2 X_1 and X_2 are independent random variables with distributions N(50, 16) and N(40, 9) respectively. Write down the distributions of

(i) $X_1 + X_2$ (ii) $X_1 - X_2$ (iii) $X_2 - X_1$.

3 A play is enjoying a long run at a theatre. It is found that the playing time may be modelled as a normal variable with mean 130 minutes and standard deviation 3 minutes, and that the length of the intermission in the middle of the performance may be modelled by a normal variable with mean 15 minutes and standard deviation 5 minutes. Find the probability that the performance is completed in less than 140 minutes.

4 The time Melanie spends on her history assignments may be modelled as being normally distributed, with mean 40 minutes and standard deviation 10 minutes. The times taken on assignments may be assumed to be independent. Find

  (i) the probability that a particular assignment will last longer than an hour
  (ii) the time in which 95% of all assignments can be completed
  (iii) the probability that two assignments will be completed in less than 75 minutes.

5 The weights of full cans of a particular brand of pet food may be taken to be normally distributed, with mean 260 g and standard deviation 10 g. The weights of the empty cans may be taken to be normally distributed, with mean 30 g and standard deviation 2 g. Find

  (i) the mean and standard deviation of the weights of the contents of the cans
  (ii) the probability that a full can weighs more than 270 g
  (iii) the probability that two full cans together weigh more than 540 g.

6 The independent random variables X_1 and X_2 are distributed as follows:

$$X_1 \sim N(30, 9); \quad X_2 \sim N(40, 16).$$

Find the distributions of the following:

  (i) $X_1 + X_2$
  (ii) $X_1 - X_2$.

7 In a vending machine the capacity of cups is normally distributed, with mean 200 cm^3 and standard deviation 4 cm^3. The volume of coffee discharged per cup is normally distributed, with mean 190 cm^3 and standard deviation 5 cm^3. Find the percentage of drinks which overflow.

8 On a distant island the heights of adult men and women may both be taken to be normally distributed, with means 173 cm and 165 cm and standard deviations 10 cm and 8 cm respectively.

  (i) Find the probability that a randomly chosen woman is taller than a randomly chosen man.
  (ii) Do you think that this is equivalent to the probability that a married woman is taller than her husband?

9 The lifetimes of a certain brand of refrigerator are approximately normally distributed, with mean 2000 days and standard deviation 250 days. Mrs Chudasama and Mr Poole each buy one on the same date.

What is the probability that Mr Poole's refrigerator is still working one year after Mrs Chudasama's refrigerator has broken down?

10 A random sample of size 2 is chosen from a normal distribution N(100, 10). Find the probability that

  (i) the sum of the sample numbers exceeds 225
  (ii) the first observation is at least 12 more than the second observation.

More than two independent random variables

The results on page 56 may be generalised to give the mean and variance of the sums and differences of n random variables, $X_1, X_2, \ldots, X_n$.

- $E(X_1 \pm X_2 \pm \ldots \pm X_n) = E(X_1) \pm E(X_2) \pm \ldots \pm E(X_n)$

and, provided $X_1, X_2, \ldots, X_n$ are independent,

- $\text{Var}(X_1 \pm X_2 \pm \ldots \pm X_n) = \text{Var}(X_1) + \text{Var}(X_2) + \ldots + \text{Var}(X_n)$.

If $X_1, X_2, \ldots, X_n$ is a set of normally distributed variables, then the distribution of $(X_1 \pm X_2 \pm \ldots \pm X_n)$ is also normal.

EXAMPLE 2.4

The mass, X, of a suitcase at an airport is modelled as being normally distributed, with mean 15 kg and standard deviation 3 kg. Find the probability that a random sample of ten suitcases weighs more than 154 kg.

SOLUTION

The mass X of one suitcase is given by

$$X \sim N(15, 9).$$

Then the mass of each of the ten suitcases has the distribution of X; call them $X_1, X_2, \ldots, X_{10}$.

Let the random variable T be the total weight of ten suitcases.

$$T = X_1 + X_2 + \ldots + X_{10}.$$
$$E(T) = E(X_1) + E(X_2) + \ldots + E(X_{10})$$
$$= 15 + 15 + \ldots + 15$$
$$= 150$$

Similarly
$$\text{Var}(T) = \text{Var}(X_1) + \text{Var}(X_2) + \ldots + \text{Var}(X_{10})$$
$$= 9 + 9 + \ldots + 9$$
$$= 90$$

So $T \sim N(150, 90)$

The probability that T exceeds 154 is given by

$$1 - \Phi\left(\frac{154 - 150}{\sqrt{90}}\right)$$
$$= 1 - \Phi(0.422)$$
$$= 1 - 0.6635$$
$$= 0.3365$$

standard deviation = $\sqrt{90}$

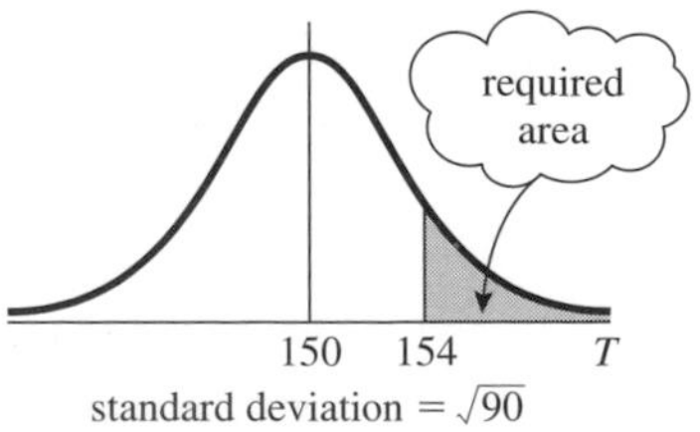

Figure 2.5

EXAMPLE 2.5

The running times of the four members of a $4 \times 400\,\text{m}$ relay race may all be taken to be normally distributed, as follows:

Member	Mean time (s)	Standard deviation (s)
Adil	52	1
Brian	53	1
Colin	55	1.5
Dexter	51	0.5

Assuming that no time is lost during changeovers, find the probability that the team finishes the race in less than 3 minutes 28 seconds.

SOLUTION

Let the total time be T.

$$\text{E}(T) = 52 + 53 + 55 + 51 = 211$$
$$\text{Var}(T) = 1^2 + 1^2 + 1.5^2 + 0.5^2$$
$$= 1 + 1 + 2.25 + 0.25 = 4.5$$

So $T \sim \text{N}(211, 4.5)$.

The probability of a total time of less than 3 minutes 28 seconds (208 seconds) is given by

$$\Phi\left(\frac{208 - 211}{\sqrt{4.5}}\right) = \Phi(-1.414)$$
$$= 1 - 0.9213$$
$$= 0.0787$$

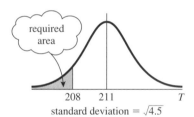

Figure 2.6

Linear combinations of two or more independent random variables

The results given on page 56 can also be generalised to include linear combinations of random variables.

For any random variables X and Y,

- $\text{E}(aX + bY) = a\text{E}(X) + b\text{E}(Y)$, where a and b are constants.

If X and Y are independent

- $\text{Var}(aX + bY) = a^2\text{Var}(X) + b^2\text{Var}(Y)$.

If the distributions of X and Y are normal, then the distribution of $(aX + bY)$ is also normal.

These results may be extended to any number of random variables.

EXAMPLE 2.6

In a workshop joiners cut out rectangular sheets of laminated board, of length L cm and width W cm, to be made into work surfaces. Both L and W may be taken to be normally distributed with standard deviation 1.5 cm. The mean of L is 150 cm, that of W is 60 cm, and the lengths of L and W are independent. Both of the short sides and one of the long sides have to be covered by a protective strip (the other long side is to lie against a wall and so does not need protection).

What is the probability that a protecting strip 275 cm long will be too short for a randomly selected work surface?

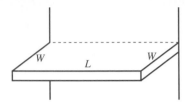

Figure 2.7

SOLUTION

Denoting the length and width by the independent random variables L and W and the total length of strip required by T:

$$T = L + 2W$$
$$E(T) = E(L) + 2E(W)$$
$$= 150 + 2 \times 60$$
$$= 270$$
$$\mathrm{Var}(T) = \mathrm{Var}(L) + 2^2\mathrm{Var}(W)$$
$$= 1.5^2 + 4 \times 1.5^2$$
$$= 11.25$$

The probability of a strip 275 cm long being too short is given by

$$1 - \Phi\left(\frac{275 - 270}{\sqrt{11.25}}\right) = 1 - \Phi(1.491)$$
$$= 1 - 0.932$$
$$= 0.068.$$

Note

You have to distinguish carefully between the random variable $2W$, which means twice the size of one observation of the random variable W, and the random variable $W_1 + W_2$, which is the sum of two independent observations of the random variable W.

In the last example $\quad\quad\quad E(2W) = 2E(W) = 120$

and $\quad\quad\quad\quad\quad\quad \mathrm{Var}(2W) = 2^2\mathrm{Var}(W) = 4 \times 2.25 = 9.$

In contrast, $\quad\quad\quad E(W_1 + W_2) = E(W_1) + E(W_2) = 60 + 60 = 120$

and $\quad\quad\quad\quad\quad\quad \mathrm{Var}(W_1 + W_2) = \mathrm{Var}(W_1) + \mathrm{Var}(W_2) = 2.25 + 2.25 = 4.5.$

EXAMPLE 2.7

A machine produces sheets of paper the thicknesses of which are normally distributed with mean 0.1 mm and standard deviation 0.006 mm.

(i) State the distribution of the total thickness of eight randomly selected sheets of paper.

(ii) Single sheets of paper are folded three times (to give eight thicknesses). State the distribution of the total thickness.

SOLUTION

Denote the thickness of one sheet (in mm) by the random variable W, and the total thickness of eight sheets by T.

(i) *Eight separate sheets*

In this situation $T = W_1 + W_2 + W_3 + W_4 + W_5 + W_6 + W_7 + W_8$

where $W_1, W_2, \ldots, W_8$ are eight independent observations of the variable W. The distribution of W is normal with mean 0.1 and variance 0.006^2.

So the distribution of T is normal with

$$\text{mean} = 0.1 + 0.1 + \ldots + 0.1 = 8 \times 0.1 = 0.8$$

$$\text{variance} = 0.006^2 + 0.006^2 + \ldots + 0.006^2 = 8 \times 0.006^2$$

$$= 0.000\,288$$

$$\text{standard deviation} = \sqrt{0.000\,288} = 0.017.$$

The distribution is $N(0.8, 0.017^2)$.

(ii) *Eight thicknesses of the same sheet*

In this situation $T = W_1 + W_1 + W_1 + W_1 + W_1 + W_1 + W_1 + W_1 = 8W_1$ where W_1 is a single observation of the variable W.

So the distribution of T is normal with

$$\text{mean} = 8 \times E(W) = 0.8$$

$$\text{variance} = 8^2 \times \text{Var}(W) = 8^2 \times 0.006^2 = 0.002\,304$$

$$\text{standard deviation} = \sqrt{0.002\,304} = 0.048.$$

The distribution is $N(0.8, 0.048^2)$.

? Notice that in both cases the mean thickness is the same but for the folded paper the variance is greater. Why is this?

EXERCISE 2B

1 A garage offers motorists 'MOT While U Wait' and claims that an average test takes only 20 minutes. Assuming that the time taken can be modelled as a normal variable with mean 20 minutes and standard deviation 2 minutes, find the distribution of the total time taken to conduct 6 MOTs in succession at this garage. State any assumptions you make.

2 A company manufactures floor tiles of mean length 20 cm with standard deviation 0.2 cm. Assuming the distribution of the lengths of the tiles is normal, find the probability that, when 12 randomly selected floor tiles are laid in a row, their total length exceeds 241 cm.

3 The masses of Christmas cakes produced at a bakery are independent and may be modelled as being normally distributed with mean 4 kg and standard deviation 100 g. Find the probability that a set of eight Christmas cakes has a total mass between 32.3 kg and 32.7 kg.

4 A random sample of 15 items is chosen from a normal population with mean 30 and variance 9. Find the probability that the sum of the variables in the sample is less than 440.

5 The distributions of four independent random variables X_1, X_2, X_3 and X_4 are N(7, 9), N(8, 16), N(9, 4) and N(10, 1) respectively. Find the distributions of
(i) $X_1 + X_2 + X_3 + X_4$ **(ii)** $X_1 + X_2 - X_3 - X_4$ **(iii)** $X_1 + X_2 + X_3$.

6 The distributions of X and Y are N(100, 25) and N(110, 36), and X and Y are independent. Find
(i) the probability that $8X + 2Y < 1000$
(ii) the probability that $8X - 2Y > 600$.

7 The distributions of the independent random variables A, B and C are N(35, 9), N(30, 8) and N(35, 9). Write down the distributions of
(i) $A + B + C$ **(ii)** $5A + 4B$ **(iii)** $A + 2B + 3C$ **(iv)** $4A - B - 5C$.

8 The distributions of the independent random variables X and Y are N(60, 4) and N(90, 9). Find the probability that
(i) $X - Y < -35$ **(ii)** $3X + 5Y > 638$ **(iii)** $3X > 2Y$.

9 If $X \sim$ N(60, 4) and $Y \sim$ N(90, 9) and X and Y are independent, find the probability that
(i) when one item is sampled from each population, the one from the Y population is more than 35 greater than the one from the X population
(ii) the sum of a sample consisting of three items from population X and five items from population Y exceeds 638
(iii) the sum of a sample of three items from population X exceeds that of two items from population Y.
(iv) Comment on your answers to questions 8 and 9.

10 If $X_1 \sim$ N(600, 400) and $X_2 \sim$ N(1000, 900) and X_1 and X_2 are independent, write down the distributions of
(i) $4X_1 + 5X_2$
(ii) $7X_1 - 3X_2$
(iii) $aX_1 + bX_2$, where a and b are constants.

11 The distribution of the weights of those rowing in a very large regatta may be taken to be normal with mean 80 kg and standard deviation 8 kg.

    (i) What total weight would you expect 70% of randomly chosen crews of four oarsmen to exceed?

    (ii) State what assumption you have made in answering this question and comment on whether you consider it reasonable.

12 The quantity of fuel used by a coach on a return trip of 200 km is modelled as a normal variable with mean 45 l and standard deviation 1.5 l. Find

    (i) the probability that in nine return journeys the coach uses between 400 and 406 l of fuel

    (ii) the volume of fuel which is 95% certain to be sufficient to cover the total fuel requirements for two return journeys.

13 A doctor has found that the times taken to examine patients coming into the surgery are independent and normally distributed with a mean of 9 minutes and a standard deviation of 3 minutes. The doctor sees patients consecutively with no time gaps between them. On a particular day there are 16 patients in the surgery and the doctor sees the first patient at 9.00 am.

    (i) Find the probability that exactly two of the first six patients have examination times in excess of 9 minutes.

    (ii) Find the probability that the doctor will have examined all 16 patients by 11.30 am.

[WJEC]

14 Assume that the weights of men and women may be taken to be normally distributed, men with mean 75 kg and standard deviation 4 kg, and women with mean 65 kg and standard deviation 3 kg.

At a village fair, tug-of-war teams consisting of either five men or six women are chosen at random. The competition is then run on a knock-out basis, with teams drawn out of a hat. If in the first round a women's team is drawn against a men's team, what is the probability that the women's team is the heavier? State any assumptions you have made and explain how they can be justified.

15 A school student investigated how long he actually had to spend on homework assignments, which were nominally for half-hour periods. He found that the times were approximately normally distributed, with mean 35 minutes and standard deviation 8 minutes. Using this model, and assuming independence between assignments, find

    (i) the probability that three assignments each take more than 40 minutes

    (ii) the probability that three assignments will take more than 2 hours altogether.

[Cambridge]

16 The length, in centimetres, of a rectangular tile is a normal variable with mean 19.8 and standard deviation 0.1. The breadth, in centimetres, is an independent normal variable with mean 9.8 and standard deviation 0.1.

(i) Find the probability that the sum of the lengths of five randomly chosen tiles exceeds 99.4 cm.

(ii) Find the probability that the breadth of a randomly chosen tile is less than one half of the length.

(iii) S denotes the sum of the lengths of 50 randomly chosen tiles and T denotes the sum of the breadths of 90 randomly chosen tiles. Find the mean and variance of $S - T$.

[Cambridge]

17 The weights of pamphlets are normally distributed with mean 40 g and standard deviation 2 g. What is the distribution of the total weight of

(i) a random sample of 2 pamphlets?

(ii) a random sample of n pamphlets?

Pamphlets are stacked in piles nominally containing 25. To save time, the following method of counting is used. A pile of pamphlets is weighed and is accepted (i.e. assumed to contain 25 pamphlets) if its weight lies between 980 g and 1020 g. Assuming each pile is a random sample, determine to three decimal places the probabilities that

(iii) a pile actually containing 24 pamphlets will be accepted

(iv) a pile actually containing 25 pamphlets will be rejected.

Justify the choice of the limits as 980 g and 1020 g.

[MEI]

18 The four runners in a relay team have individual times, in seconds, which are normally distributed, with means 12.1, 12.2, 12.3, 12.4, and standard deviations 0.2, 0.25, 0.3, 0.35 respectively. Find the probability that, in a randomly chosen race,

(i) the total time of the four runners is less than 48 seconds

(ii) runners 1 and 2 take longer in total than do runners 3 and 4.

What assumption have you made and how realistic is the model?

19 A petrol company issues a voucher with every 12 litres of petrol that a customer buys. Customers who send 50 vouchers to Head Office are entitled to a 'free gift'. After this promotion has been running some time the company receives several hundred bundles of vouchers in each day's post. It would take a long time, and so be costly, for somebody to count each bundle and so they weigh them instead. The weight of a single voucher is a normal variable with mean 40 mg and standard deviation 5 mg.

(i) What is the distribution of the weights of bundles of 50 vouchers?

(ii) Find the weight W mg which is exceeded by 95% of bundles.

The company decides to count only the number of vouchers in those bundles which weigh less than W mg. A man has only 48 vouchers but decides to send them in, claiming that there are 50.

(iii) What is the probability that the man is detected?

20 Jim Longlegs is an athlete whose specialist event is the triple jump. This is made up of a *hop*, a *step* and a *jump*. Over a season the lengths of the *hop*, *step* and *jump* sections, denoted by H, S and J respectively, are measured, from which the following models are proposed:

$$H \sim N(5.5, 0.5^2) \quad S \sim N(5.1, 0.6^2) \quad J \sim N(6.2, 0.8^2)$$

where all distances are in metres. Assume that H, S and J are independent.

(i) In what proportion of his triple jumps will Jim's total distance exceed 18 metres?

(ii) In six successive independent attempts, what is the probability that at least one total distance will exceed 18 m?

(iii) What total distance will Jim exceed 95% of the time?

(iv) Find the probability that, in Jim's next triple jump, his step will be greater than his hop.

[MEI]

21 A country baker makes biscuits whose masses are normally distributed with mean 30 g and standard deviation 2.3 g. She packs them by hand into either a small carton (containing 20 biscuits) or a large carton (containing 30 biscuits).

(i) State the distribution of the total mass, S, of biscuits in a small carton and find the probability that S is greater than 615 g.

(ii) Six small and four large cartons are placed in a box. Find the probability that the total mass of biscuits in the ten cartons lies between 7150 g and 7250 g.

(iii) Find the probability that three small cartons contain at least 25 g more than two large ones.

The label on a large carton of biscuits reads 'Net mass 900 g'. A trading standards officer insists that 90% of such cartons should contain biscuits with a total mass of at least 900 g.

(iv) Assuming the standard deviation remains unchanged, find the least value of the mean mass of a biscuit consistent with this requirement.

[MEI]

22 The continuous random variables X and Y represent the masses of male and female students who attend my local college.

Both X and Y are normally distributed such that $X \sim N(75, 6^2)$ and $Y \sim N(65, 5^2)$, where all masses are given in kilograms.

(i) Find the probability that, if a male student and a female student are chosen at random, they each have a mass exceeding 70 kg.

(ii) State carefully the distribution of the combined mass of a random sample of m male and f female students.

A lift in the college has a notice

> **MAXIMUM 8 PEOPLE** *or* **650 kg**

Find the probability that the combined mass of a random sample of eight students will exceed the mass restriction if it consists of

(a) eight males

(b) five males and three females.

(iii) What is the probability that a randomly selected female student has a greater mass than a randomly selected male student?

[MEI]

23 During the hour from 10.00 am to 11.00 am on a Wednesday morning, customers enter a large store at a constant average rate of 6.2 per minute.

(i) Use a suitable distribution to find the probability that more than six customers enter the store in any given minute. What assumption have you made about customers' behaviour?

(ii) Give a suitable approximating distribution for the total number of customers entering during the hour. Use this distribution to find the probability that more than 350 customers enter the store during the hour.

Suppose now that during this hour customers leave the store at a constant average rate of 5.7 per minute.

(iii) Obtain a distribution for the increase in the number of customers during the hour. Use this distribution to find the probability that more customers leave than enter during the hour.

[MEI]

24 A computer is used to add up a series of numbers. Each addition introduces an error which may be regarded as a random variable, X, which has the rectangular distribution on the interval $[-a, a]$, where a is, of course, very small. The probability density function for X is illustrated below.

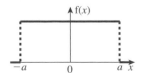

Find, in terms of a,

(i) $f(x)$, the probability density function for X

(ii) $E(X)$ and $Var(X)$.

Now suppose that the total number of additions performed is n, and that successive errors are independent. Let the total error be Y.

(iii) Write down the values of $E(Y)$ and $Var(Y)$ in terms of a.

(iv) For the case $a = 5.0 \times 10^{-11}$ and $n = 1000$ state the greatest possible value of the total error, Y.

The distribution of the sample mean

In many practical situations you do not know the true value of the mean of a variable that you are investigating, that is the parent population mean (usually just called the *population mean*). Indeed that may be one of the things you are trying to establish.

In such cases you will usually take a random sample, $x_1, x_2, \ldots, x_n$, of size n from the population and work out the sample mean $\bar{x}$,

$$\bar{x} = \frac{x_1 + x_2 + \ldots + x_n}{n}$$

to use as an estimate for the true population mean μ.

How accurate is this estimate likely to be and how does its reliability vary with n, the sample size?

Each of the sample values $x_1, x_2, \ldots, x_n$ can be thought of as a value of an independent random variable $X_1, X_2, \ldots, X_n$. The variables $X_1, X_2, \ldots, X_n$ have the same distribution as the population and so $E(X_1) = \mu$, $Var(X_1) = \sigma^2$, etc.

So the sample mean is a value of the random variable $\bar{X}$ given by

$$\bar{X} = \frac{1}{n}(X_1 + X_2 + \ldots + X_n)$$

$$= \frac{1}{n}X_1 + \frac{1}{n}X_2 + \ldots + \frac{1}{n}X_n$$

and so

$$E(\bar{X}) = \frac{1}{n}E(X_1) + \frac{1}{n}E(X_2) + \ldots + \frac{1}{n}E(X_n)$$

$$= \frac{1}{n}\mu + \frac{1}{n}\mu + \ldots + \frac{1}{n}\mu$$

$$= \mu, \text{ the population mean.}$$

Further, using the fact that $X_1, X_2, \ldots, X_n$ are independent,

$$\mathrm{Var}(\overline{X}) = \mathrm{Var}\left(\frac{X_1}{n}\right) + \mathrm{Var}\left(\frac{X_2}{n}\right) + \ldots + \mathrm{Var}\left(\frac{X_n}{n}\right)$$

$$= \frac{1}{n^2}\mathrm{Var}(X_1) + \frac{1}{n^2}\mathrm{Var}(X_2) + \ldots + \frac{1}{n^2}\mathrm{Var}(X_n)$$

$$= \frac{1}{n^2}\sigma^2 + \frac{1}{n^2}\sigma^2 + \ldots + \frac{1}{n^2}\sigma^2$$

$$= n\left(\frac{1}{n^2}\sigma^2\right)$$

$$= \frac{\sigma^2}{n}$$

Thus the distribution of the means of samples of size n, drawn from a parent population with mean μ and variance σ^2, has mean μ and variance $\frac{\sigma^2}{n}$. The distribution of the sample means is called the *sampling distribution of the means,* or just the *sampling distribution.*

Notice that $\mathrm{Var}(\overline{X}) = \frac{\sigma^2}{n}$ means that as n increases the variance of the sample means decreases. In other words the value obtained for $\overline{X}$ from a large sample is more reliable as an estimate for μ than one obtained from a smaller sample. This result, simple though it is, lies at the heart of statistics: it says that you are likely to get more accurate results if you take a larger sample.

The standard deviation of sample means of size n is $\frac{\sigma}{\sqrt{n}}$ and this is called the *standard error of the mean,* or often just the *standard error.* It gives a measure of the degree of accuracy of $\overline{X}$ as an estimate for μ.

Note

The derivation has required no assumptions about the distribution of the parent population, other than that μ and σ are finite. If in fact the parent distribution is normal, then the sampling distribution will also be normal, whatever the size of n.

If the parent population is not normal, the sampling distribution will still be approximately normal, and will be more accurately so for larger values of n.

This result is called the central limit theorem and will be developed in chapter 3.

The derivation does require that the sample items are independent (otherwise the result for $\mathrm{Var}(X)$ would not have been valid).

EXAMPLE 2.8

2

The distribution of the sample mean

The discrete random variable X has a probability distribution as shown:

X	1	2	3
Probability	0.5	0.4	0.1

A random sample of size 2 is chosen, with replacement after each selection.
(i) Find μ and σ^2.
(ii) Verify that $E(\bar{X}) = \mu$ and $Var(\bar{X}) = \dfrac{\sigma^2}{2}$.

SOLUTION

(i)
$$\mu = E(X) = 1 \times 0.5 + 2 \times 0.4 + 3 \times 0.1$$
$$= 1.6$$
$$E(X^2) = 1^2 \times 0.5 + 2^2 \times 0.4 + 3^2 \times 0.1$$
$$= 3$$
$$\sigma^2 = Var(X) = E(X^2) - [E(X)]^2$$
$$= 3 - 1.6^2 = 0.44$$

(ii) The table below lists all the possible samples of size 2, their means and probabilities.

Sample	1, 1	1, 2	1, 3	2, 1	2, 2	2, 3	3, 1	3, 2	3, 3
Mean	1	1.5	2	1.5	2	2.5	2	2.5	3
Probability	0.25	0.2	0.05	0.2	0.16	0.04	0.05	0.04	0.01

This gives the following probability distribution of the sample mean:

$\bar{X}$	1	1.5	2	2.5	3
Probability	0.25	0.4	0.26	0.08	0.01

Using this table gives

$$E(\bar{X}) = 1 \times 0.25 + 1.5 \times 0.4 + 2 \times 0.26 + 2.5 \times 0.08 + 3 \times 0.01$$
$$= 1.6 = \mu, \text{ as required.}$$
$$E(\bar{X}^2) = 1^2 \times 0.25 + 1.5^2 \times 0.4 + 2^2 \times 0.26 + 2.5^2 \times 0.08 + 3^2 \times 0.01$$
$$= 2.78$$
$$Var(\bar{X}) = E(\bar{X}^2) - [E(\bar{X})]^2$$
$$= 2.78 - (1.6)^2$$
$$= 0.22$$
$$= \frac{0.44}{2} = \frac{\sigma^2}{2}, \text{ as required.}$$

The χ^2 (chi squared) distribution

You have seen that if n independent normal variables, $X_1, X_2, X_3, \ldots, X_n$ are added together, the resulting variable

$$T = X_1 + X_2 + X_3 + \ldots + X_n$$

is itself normal. If the normal variables have been standardised so that each has mean 0 and variance 1, then the variable T has mean 0 and variance n.

A different, but important, distribution is formed when the squares of a set of independent standardised normal variables are added. This is the χ^2 (chi squared) distribution which is used in the χ^2 test (the subject of Chapter 5 of this book) which evaluates how well a model fits data.

If $U = Z_1^2 + Z_2^2 + Z_3^2 + \ldots + Z_n^2$

then $U \sim \chi_n^2$

where n denotes the number of independent random variables involved, which is called the *degrees of freedom*. The Greek letter v (nu) is usually used to denote the degrees of freedom.

Figure 2.8 shows the p.d.f.s of the χ^2 distribution for various values of v.

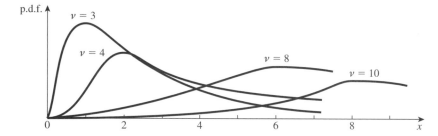

Figure 2.8

Note

As can be seen from figure 2.8, the shapes of the curves are alike and are all skewed. For large values of v they can be well approximated by the distribution $N(v, 2v)$.

The graphs of the χ^2 distribution in the cases $v = 1$ and $v = 2$ are quite different from all the others and are shown in figure 2.9.

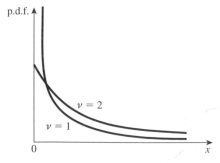

Figure 2.9

1 For two random variables X and Y, whether independent or not, and constants a and b,

- $E(X \pm Y) = E(X) \pm E(Y)$
- $E(aX + bY) = aE(X) + bE(Y)$

and, if X and Y are independent,

- $\text{Var}(X \pm Y) = \text{Var}(X) + \text{Var}(Y)$
- $\text{Var}(aX + bY) = a^2\text{Var}(X) + b^2\text{Var}(Y)$

2 For a set of n random variables, $X_1, X_2, \ldots, X_n$

- $E(X_1 \pm X_2 \pm \ldots \pm X_n) = E(X_1) \pm E(X_2) \pm \ldots \pm E(X_n)$

and, if the variables are independent,

- $\text{Var}(X_1 \pm X_2 \pm \ldots \pm X_n) = \text{Var}(X_1) + \text{Var}(X_2) + \ldots + \text{Var}(X_n)$

3 If random variables are normally distributed so are the sums, differences and other linear combinations of them.

4 **The sampling distribution of the means**

For samples of size n drawn from an infinite, or large, population with mean μ and variance σ^2, or for sampling with replacement,

- $E(\bar{X}) = \mu$
- $\text{Var}(\bar{X}) = \dfrac{\sigma^2}{n}$ where $\bar{X}$ is the sample mean.

The standard deviation of the sample means, $\frac{\sigma}{\sqrt{n}}$, is called the standard error of the mean.

5 **The χ^2 (chi squared) distribution**

If $U = Z_1{}^2 + Z_2{}^2 + \ldots + Z_n{}^2$

where $Z_1{}^2, Z_2{}^2, Z_3{}^2$, etc. are the squares of independent standardised normal variables, then

$$U \sim \chi_n^2$$

Interpreting sample data using the normal distribution

When we spend money on testing an item, we are buying confidence in its performance.

Tony Cutler

THE AVONFORD STAR

Avonford set to become greenhouse?

From our Science Correspondent Ama Williams

On a recent visit to Avonford Community College, I was intrigued to find experiments being conducted to measure the level of carbon dioxide in the air we are all breathing. Readers will of course know that high levels of carbon dioxide are associated with the greenhouse effect.

Lecturer Ray Sharp showed me round his laboratory. 'It is delicate work, measuring parts per million, but I am trying to establish what is the normal level in this area. Yesterday we took ten readings and you can see the results for yourself: 336, 334, 332, 332, 331, 331 330, 330, 328, 326.'

When I commented that there seemed to be a lot of variation between the readings, Ray assured me that that was quite in order.

'I have taken hundreds of these measurements in the past,' he said. 'There is always a standard deviation of 2.5. That's just natural variation.'

I suggested to Ray that his students should test whether these results are significantly above

the accepted value of 328 parts per million. Certainly they made me feel uneasy. Is the greenhouse effect starting here in Avonford?

Ray Sharp has been trying to establish the carbon dioxide level at Avonford. How do you interpret his figures? Do you think the correspondent has a point when she says she is worried that the greenhouse effect is already happening in Avonford?

If suitable sampling procedures have not been used, then the resulting data may be worthless, indeed positively misleading. You may wonder if that is the case with Ray's figures, and about the accuracy of his analysis of the samples too. His data are used in subsequent working in this chapter, but you may well feel there is something of a question mark hanging over them. You should always be prepared to treat data with a healthy degree of caution.

Putting aside any concerns about the quality of the data, what conclusions can you draw from them?

Estimating the population mean, μ

Ray Sharp's data were as follows:

$$336, 334, 332, 332, 331, 331, 330, 330, 328, 326$$

His intention in collecting them was to estimate the mean of the parent population, the population mean.

The mean of these figures, the sample mean, is given by

$$\bar{x} = \frac{(336 + 334 + 332 + 332 + 331 + 331 + 330 + 330 + 328 + 326)}{10}$$

$$= 331$$

What does this tell you about the population mean, μ?

It tells you that it is about 331 but it certainly does not tell you that it is definitely and exactly 331. If Roy took another sample, its mean would probably not be 331 but you would be surprised (and suspicious) if it were very far away from it. If he took lots of samples, all of size 10, you would expect their means to be close together but certainly not all the same.

If you took 1000 such samples, each of size 10, the distribution of their means might look like figure 3.1.

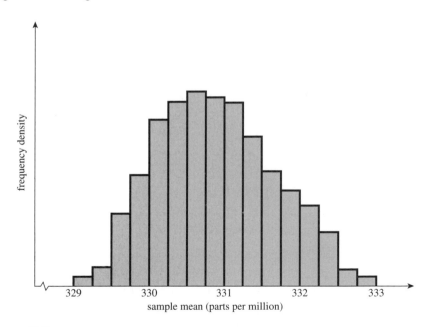

Figure 3.1

You will notice that this distribution looks rather like the normal distribution and so may well wonder if this is indeed the case. The answer, provided by the central limit theorem, is yes.

The central limit theorem

For samples of size n drawn from a distribution with mean μ and finite variance σ^2, the distribution of the sample mean is approximately $N\left(\mu, \dfrac{\sigma^2}{n}\right)$ for sufficiently large n.

This theorem is fundamental to much of statistics and so it is worth pausing to make sure you understand just what it is saying.

It deals with the distribution of sample means. This is called the *sampling distribution* (or more correctly the *sampling distribution of the means*). There are three aspects to it.

1 The mean of the sample means is μ, the population mean of the original distribution. That is not a particularly surprising result but it is extremely important.

2 The standard deviation of the sample means is $\dfrac{\sigma}{\sqrt{n}}$. This is often called the *standard error of the mean*.

Within a sample you would expect some values above the population mean, others below it, so that overall the deviations would tend to cancel each other out, and the larger the sample the more this would be the case. Consequently the standard deviation of the sample means is smaller than that of individual items, by a factor of $\sqrt{n}$.

3 The distribution of sample means is approximately normal.

This last point is the most surprising part of the theorem. Even if the underlying parent distribution is not normal, the distribution of the means of samples of a particular size drawn from it is approximately normal. The larger the sample size, n, the closer this distribution is to the normal. For any given value of n the sampling distribution will be closest to normal where the parent distribution is not unlike the normal.

In many cases the value of n does not need to be particularly large. For most parent distributions you can rely on the distribution of sample means being normal if n is about 20 or 25 (or more).

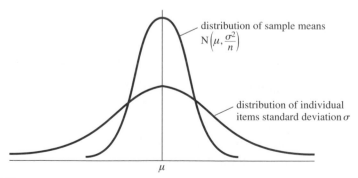

distribution of sample means $N\left(\mu, \dfrac{\sigma^2}{n}\right)$

distribution of individual items standard deviation σ

μ

Figure 3.2

Confidence intervals

Returning to the figures on carbon dioxide levels, you would estimate the population mean to be the same as the sample mean, namely 331.

You could express this by saying that you estimate μ to lie within a range of values, an interval, centred on 331

$$331 - \text{a bit} < \mu < 331 + \text{a bit.}$$

Such an interval is called a *confidence interval.*

Imagine you take a large number of samples and use a formula to work out the interval for each of them. If you catch the true population mean in 90% of your intervals, the confidence interval is called a 90% confidence interval. Other percentages are also used and the confidence intervals are named accordingly. The width of the interval is clearly twice the 'bit'.

Finding a confidence interval involves a very simple calculation but the reasoning behind it is somewhat subtle and requires clear thinking. It is explained in the next section, but you may prefer to make your first reading of it a light one. You should, however, come back to it at some point; otherwise you will not really understand the meaning of confidence intervals.

Theory of confidence intervals

To understand confidence intervals you need to look not at the particular sample whose mean you have just found, but at the parent population from which it was drawn. For Ray Sharpe's situation this does not look very promising. All you know about it is its standard deviation σ (in this case 2.5). You do not know its mean, μ, which you are trying to estimate, or even its shape.

It is now that the strength of the central limit theorem becomes apparent. This states that the distribution of the means of samples of size n drawn from this population is approximately normal with mean μ and standard deviation $\dfrac{\sigma}{\sqrt{n}}$.

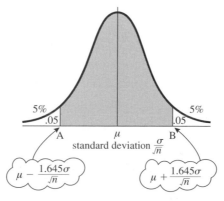

Figure 3.3

In figure 3.3 the central 90% region has been shaded leaving the two 5% tails, corresponding to z values of ±1.645, unshaded. So if you take a large number of samples, all of size n, and work out the sample mean $\bar{x}$ for each one, you would expect that in 90% of cases the value of $\bar{x}$ would lie in the shaded region between A and B.

For such a value of $\bar{x}$ to be in the shaded region

it must be to the right of A: $\qquad\qquad \bar{x} > \mu - 1.645\dfrac{\sigma}{\sqrt{n}}$ ①

it must be to the left of B: $\qquad\qquad \bar{x} < \mu + 1.645\dfrac{\sigma}{\sqrt{n}}$ ②

Rearranging these two inequalities:

$$\text{①} \qquad \bar{x} + 1.645\dfrac{\sigma}{\sqrt{n}} > \mu \qquad \text{or } \mu < \bar{x} + 1.645\dfrac{\sigma}{\sqrt{n}}$$

$$\text{②} \qquad \bar{x} - 1.645\dfrac{\sigma}{\sqrt{n}} < \mu$$

Putting them together gives the result that in 90% of cases

$$\bar{x} - 1.645\dfrac{\sigma}{\sqrt{n}} < \mu < \bar{x} + 1.645\dfrac{\sigma}{\sqrt{n}}$$

and this is the 90% confidence interval for μ.

The numbers corresponding to the points A and B are called the 90% *confidence limits* and 90% is the *confidence level*.

If you want a different confidence level, you use a different z value from 1.645.

This number is often denoted by k; commonly used values are:

Confidence level	k
90%	1.645
95%	1.96
99%	2.58

and the confidence interval is given by

$$\bar{x} - k\dfrac{\sigma}{\sqrt{n}} \quad \text{to} \quad \bar{x} + k\dfrac{\sigma}{\sqrt{n}}.$$

The P% confidence interval for the mean is an interval constructed from sample data in such a way that P% of such intervals will include the true population mean. Figure 3.4 shows a number of confidence intervals constructed from different samples, one of which fails to catch the population mean.

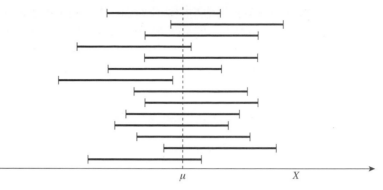

Figure 3.4

Note

Notice that this is a two-sided symmetrical confidence interval for the mean, μ. Confidence intervals do not need to be symmetrical and can be one-sided. The term confidence interval is a general one, applying not just to the mean but to other population parameters, like variance and skew, as well. All these cases, however, are outside the scope of this book.

In the case of Ray Sharp's data,

$$\bar{x} = 331, \quad \sigma = 2.5, \quad n = 10$$

and so the 90% confidence interval is

$$331 - 1.645 \times \frac{2.5}{\sqrt{10}} \quad \text{to} \quad 331 + 1.645 \times \frac{2.5}{\sqrt{10}}$$
$$329.7 \quad \text{to} \quad 332.3$$

You will notice that the confidence interval does not include the correspondent's 'accepted value' for μ of 328. If Ray Sharp's data are reliable, the Avonford level would seem to be higher than normal.

Do you think this means that the greenhouse effect has already started in Avonford?

Known and estimated standard deviation

Notice that you can only use this procedure if you already know the value of the standard deviation of the parent population, σ. In the example, Ray Sharp had said that from taking hundreds of measurements he knew it to be 2.5.

It is more often the situation that you do not know the population standard deviation or variance, and have to estimate it from your sample data. If that is the case, the procedure is different in that you use the t-distribution rather than the normal provided that the parent population is normally distributed, and this results in different values of k. The use of the t-distribution is the subject of the next chapter.

However, if the sample is large, for example over 50, confidence intervals worked out using the normal distribution will be reasonably accurate even though the standard deviation used is an estimate from the sample. So it is quite acceptable to use the normal distribution for large samples whether the standard deviation is known or not.

EXPERIMENTS

These experiments are designed to help you understand confidence intervals, rather than to teach you anything new about dice.

When a single die is thrown, the possible outcomes, 1, 2, 3, 4, 5, 6, are all equally likely with probability $\frac{1}{6}$. Consequently the expectation or mean score from throwing a die is

$$\mu = 1 \times \tfrac{1}{6} + 2 \times \tfrac{1}{6} + \ldots + 6 \times \tfrac{1}{6} = 3.5.$$

Similarly the standard deviation is

$$\sigma = \sqrt{\left(1^2 \times \tfrac{1}{6} + 2^2 \times \tfrac{1}{6} + \ldots + 6^2 \times \tfrac{1}{6}\right) - 3.5^2} = 1.708.$$

Imagine that you know σ but don't know μ and wish to construct a 90% confidence interval for it.

CONVERGING CONFIDENCE INTERVALS

Start by throwing a die once. Suppose you get a 5. You have a sample of size 1, namely {5}, which you could use to work out a sort of 90% confidence interval (but see the warning below).

This confidence interval is given by

$$5 - 1.645 \times \frac{1.708}{\sqrt{1}} \quad \text{to} \quad 5 + 1.645 \times \frac{1.708}{\sqrt{1}}$$
$$2.19 \quad \text{to} \quad 7.81$$

 So far the procedure is not valid. The sample is small and the underlying distribution is not normal. However, things will get better. The more times you throw the die, the larger the sample size and so the more justifiable the procedure.

Now throw the die again. Suppose this time you get a 3. You now have a sample of size 2, namely {5, 3}, with mean 4, and can work out another confidence interval.

The confidence interval is given by

$$4 - 1.645 \times \frac{1.708}{\sqrt{2}} \quad \text{to} \quad 4 + 1.645 \times \frac{1.708}{\sqrt{2}}$$
$$2.79 \quad \text{to} \quad 5.21$$

Now throw the die again and find a third confidence interval, and a fourth, fifth and so on. You should find them converging on the population mean of 3.5; but it may take some time to get close, particularly if you start with, say, two 6's.

This demonstrates that, the larger the sample you take, the narrower the range of values within the confidence interval.

CATCHING THE POPULATION MEAN

Organise a group of friends to throw five dice (or one die five times), and to do this 100 times. Each of these gives a sample of size 5 and so you can use it to work out a 90% confidence interval for μ.

You know that the real value of μ is 3.5 and it should be that this is caught within 90% of 90% confidence intervals.

 Out of your 100 confidence intervals, how many actually enclose 3.5?

How large a sample do you need?

You are now in a position to start to answer the question of how large a sample needs to be. The answer, as you will see in Example 3.1, depends on the precision you require, and the confidence level you are prepared to accept.

EXAMPLE 3.1

A trading standards officer is investigating complaints that a coal merchant is giving short measure. Each sack should contain 25 kg but some variation will inevitably occur because of the size of the lumps of coal; the officer knows from experience that the standard deviation should be 1.5 kg.

The officer plans to take, secretly, a random sample of n sacks, find the total weight of the coal inside them and thereby estimate the mean weight of the coal per sack. He wants to present this figure correct to the nearest kilogram with 95% confidence. What value of n should he choose?

SOLUTION

The 95% confidence interval for the mean is given by

$$\bar{x} - 1.96 \frac{\sigma}{\sqrt{n}} \quad \text{to} \quad \bar{x} + \frac{1.96\sigma}{\sqrt{n}}$$

and so, since $\sigma = 1.5$, the inspector's requirement is that

$$\frac{1.96 \times 1.5}{\sqrt{n}} \leqslant 0.5$$

$$\frac{1.96 \times 1.5}{0.5} \leqslant \sqrt{n}$$

$$n \geqslant 34.57$$

So the inspector needs to take 35 sacks.

Large samples

Given that the width of a confidence interval decreases with sample size, why is it not standard practice to take very large samples?

The answer is that the cost and time involved has to be balanced against the quality of information produced. Because the width of a confidence interval depends on $\frac{1}{\sqrt{n}}$ and not on $\frac{1}{n}$, increasing the sample size does not produce a proportional reduction in the width of the interval. You have, for example, to increase the sample size by a factor of 4 to halve the width of the interval. In the previous example the inspector had to weigh 35 sacks of coal to achieve a class interval of $2 \times 0.5 = 1\,\text{kg}$ with 95% confidence. That is already quite a daunting task; does the benefit from reducing the interval to $0.5\,\text{kg}$ justify the time, cost and trouble involved in weighing another 105 sacks?

EXERCISE 3A

1 A biologist studying a colony of beetles selects and weighs a random sample of 20 adult males. She knows that, because of natural variability, the weights of such beetles are normally distributed with standard deviation $0.2\,\text{g}$. Their weights, in grams, are as follows:

5.2	5.4	4.9	5.0	4.8		5.7	5.2	5.2	5.4	5.1
5.6	5.0	5.2	5.1	5.3		5.2	5.1	5.3	5.2	5.2

(i) Find the mean weight of the beetles in this sample.
(ii) Find 95% confidence limits for the mean weight of such beetles.

2 An aptitude test for deep-sea divers has been designed to produce scores which are approximately normally distributed on a scale from 0 to 100 with standard deviation 25. The scores from a random sample of people taking the test were as follows:

23   35   89   35   12   45   60   78   34   66

(i) Find the mean score of the people in this sample.
(ii) Construct a 90% confidence interval for the mean score of people taking the test.
(iii) Construct a 99% confidence interval for the mean score of people taking the test. Compare this confidence interval with the 90% confidence interval.

3 In a large city the distribution of incomes per family has a standard deviation of £5200.
(i) For a random sample of 400 families, what is the probability that the sample mean income per family is within £500 of the actual mean income per family?
(ii) Given that the sample mean income was, in fact, £8300, calculate a 95% confidence interval for the actual mean income per family.

[MEI]

4 A manufacturer of women's clothing wants to know the mean height of the women in a town (in order to plan what proportion of garments should be of each size). She knows that the standard deviation of their heights is 5 cm. She selects a random sample of 50 women from the town and finds their mean height to be 165.2 cm.

(i) Use the available information to estimate the proportion of women in the town who were

(a) over 170 cm tall

(b) less than 155 cm tall.

(ii) Construct a 95% confidence interval for the mean height of women in the town.

(iii) Another manufacturer in the same town wants to know the mean height of women in the town to within 0.5 cm with 95% confidence. What is the minimum sample size that would ensure this?

5 In the table below, the masses of 40 men are recorded to the nearest kilogram.

$$
\begin{array}{cccccccc}
74 & 87 & 80 & 71 & 77 & 67 & 80 & 83 \\
78 & 84 & 75 & 79 & 73 & 79 & 81 & 77 \\
89 & 68 & 74 & 93 & 86 & 65 & 82 & 87 \\
78 & 92 & 76 & 78 & 73 & 81 & 75 & 73 \\
85 & 77 & 73 & 76 & 80 & 83 & 78 & 69
\end{array}
$$

Calculate the mean and variance of these masses.

Assuming that these masses are a random sample from a population of masses distributed normally with a variance of 40, find 95% confidence limits for the population mean. Explain carefully the meaning to be attached to these limits.

Assuming further that the population mean is 78.4, find the probability that a random sample of eight men from the population will have a total mass exceeding 640 kg.

[Cambridge]

6 An examination question, marked out of 10, is answered by a very large number of candidates. A random sample of 400 scripts are taken and the marks on this question are recorded:

Mark	0	1	2	3	4	5	6	7	8	9	10
Frequency	12	35	11	12	3	20	57	87	20	14	129

(i) Calculate the sample mean and the sample standard deviation.

(ii) Assuming that the population standard deviation has the same value as the sample standard deviation, find 90% confidence limits for the population mean.

7 An archaeologist discovers a short manuscript in an ancient language which he recognises but cannot read. There are 30 words in the manuscript and they contain a total of 198 letters. There are two written versions of the language. In the *early* form of the language the mean word length is 6.2 letters with standard deviation 2.5; in the *late* form certain words were given prefixes, raising the mean length to 7.6 letters but leaving the standard deviation unaltered. The archaeologist hopes the manuscript will help him to date the site.

(i) Construct a 95% confidence interval for the mean word length of the language.

(ii) What advice would you give the archaeologist?

8 The age, X, in years at last birthday, of 250 mothers when their first child was born is given in the following table:

Age, X	18–	20–	22–	24–	26–	28–	30–	32–	34–	36–	38–
No. of mothers	14	36	42	57	48	26	17	7	2	0	1

[The notation implies that, for example, in column 1, there are 14 mothers for whom the continuous variable X satisfies $18 \leqslant X < 20$.]

Calculate, to the nearest 0.1 of a year, estimates of the mean and the standard deviation of X.

If the 250 mothers are a random sample from a large population of mothers, find 95% confidence limits for the mean age, μ, of the total population.

[Cambridge]

9 The distribution of measurements of thicknesses of a random sample of yarns produced in a textile mill is shown in the following table.

Yarn thickness in microns (mid-interval value)	Frequency
72.5	6
77.5	18
82.5	32
87.5	57
92.5	102
97.5	51
102.5	25
107.5	9

Illustrate these data on a histogram. Estimate, to two decimal places, the mean and standard deviation of yarn thickness.

Hence estimate the standard error of the mean to two decimal places, and use it to determine approximate symmetrical 95% confidence limits, giving your answer to one decimal place.

[MEI]

10 In a game of patience, which involves no skill, the player scores between 0 and 52 points. The standard deviation is known to be 8; the mean is unknown but thought to be about 12.

   (i) Explain why players' scores cannot be normally distributed if the mean is indeed about 12.

A casino owner wishes to make this into a gambling event but needs to know the mean score before he can set the odds profitably. He employs a student to play the game 500 times. The student's total score is 6357.

   (ii) Find 99% confidence limits for the mean score.

The student recorded all her individual scores and finds on investigation that their standard deviation is not 8 but 6.21.

   (iii) What effect would accepting this value for the standard deviation have on the 99% confidence interval?

The casino owner wants to know the mean score to the nearest 0.1 with 99% confidence.

   (iv) Using the value of 6.21 for the standard deviation, find the smallest sample size that would be needed to achieve this.

Hypothesis test for the mean using the normal distribution

If your intention in collecting sample data is to test a theory, then you should set up a hypothesis test.

Ray Sharp was mainly interested in establishing data on carbon dioxide levels for Avonford. The correspondent, however, wanted to know whether levels were above normal, and so she could have set up and conducted a test.

Here is the relevant information, given in a more condensed format.

EXAMPLE 3.2   Ama Williams believes that the carbon dioxide level in Avonford has risen above the usual level of 328 parts per million. A sample of 10 specimens of Avonford air are collected and the carbon dioxide level within them is determined. The results are as follows:

$$336, 334, 332, 332, 331, 331, 330, 330, 328, 326.$$

Extensive previous research has shown that the standard deviation of the levels within such samples is 2.5, and that the distribution is near enough normal for the central limit theorem to be used for a sample of 10.

Use these data to test, at the 0.1% significance level, Ama's belief that the level of carbon dioxide at Avonford is above normal.

SOLUTION

As usual with hypothesis tests, you use the distribution of the statistic you are measuring, in this case the normal distribution of the sample means, to decide which values of the test statistic are sufficiently extreme as to suggest that the alternative hypothesis, not the null hypothesis, is true.

Null hypothesis, H_0:	$\mu = 328$	The level of carbon dioxide at Avonford is normal.
Alternative hypothesis, H_1:	$\mu > 328$	The level of carbon dioxide at Avonford is above normal.

One-tail test at the 0.1% significance level.

Method 1: Using critical regions

Since the distribution of sample means is $N\left(\mu, \dfrac{\sigma^2}{n}\right)$, critical values for a test on the sample mean are given by:

$$\mu \pm k \times \frac{\sigma}{\sqrt{n}}$$

In this case, if H_0 is true, $\mu = 328$; $\sigma = 2.5$; $n = 10$.

The test is one-tail, for $\mu > 328$, so only the right-hand tail applies. This gives a value of $k = 3.09$ since normal distribution tables give $\Phi(3.09) = 0.999$ and so $1 - \Phi(3.09) = 0.001$.

The critical value is thus $328 + 3.09 \times \dfrac{2.5}{\sqrt{10}} = 330.4$, as shown in figure 3.5.

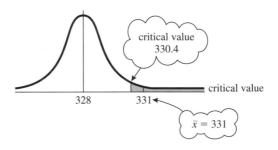

Figure 3.5

However, the sample mean $\bar{x} = 331$, and

$$331 > 330.4.$$

Therefore the sample mean lies within the critical region, and so the null hypothesis is rejected in favour of the alternative hypothesis: that the mean carbon dioxide level is above 328, at the 0.1% significance level.

Method 2: Using probabilities

The distribution of sample means, $\bar{X}$, is $N\left(\mu, \dfrac{\sigma^2}{n}\right)$.

According to the null hypothesis, $\mu = 328$ and it is known that $\sigma = 2.5$ and $n = 10$.

So this distribution is $N\left(328, \dfrac{2.5^2}{10}\right)$; see figure 3.6.

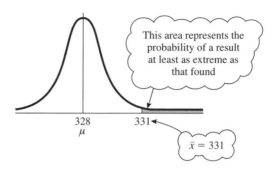

Figure 3.6

The probability of the mean, $\bar{X}$, of a randomly chosen sample being greater than the value found, i.e. 331, is given by

$$P(\bar{X} \geqslant 331) = 1 - \Phi\left(\frac{331 - 328}{\frac{2.5}{\sqrt{10}}}\right)$$

$$= 1 - \Phi(3.79)$$

$$= 1 - 0.999\,93$$

$$= 0.000\,07$$

Since $0.000\,07 < 0.001$, the required significance level (0.1%), the null hypothesis is rejected in favour of the alternative hypothesis.

Method 3: Using critical ratio

The *critical ratio* is given by $\qquad z = \dfrac{\text{observed value} - \text{expected value}}{\text{standard deviation}}$

In this case $\qquad\qquad\qquad z = \dfrac{331 - 328}{\frac{2.5}{\sqrt{10}}}$

$$= 3.79$$

This is now compared with the critical value for z, in this case $z = 3.09$. Since $\quad 3.79 > 3.09$, H_0 is rejected.

Notes

1 A hypothesis test should be formulated before the data are collected and not after. If sample data lead you to form a hypothesis, then you should plan a suitable test and collect further data on which to conduct it. It is not clear whether or not the test in the previous example was being carried out on the same data which were used to formulate the hypothesis.

2 If the data were not collected properly, any test carried out on them may be worthless.

EXAMPLE 3.3

A researcher is interested in establishing the mean IQ of the population and uses a test which is known to give scores with standard deviation 15. Initially the researcher takes a sample of 100 people and finds the mean of their scores to be 105.

(i) Calculate a 95% confidence interval for the population mean.

Knowing that 50 years earlier the mean score on this test was 100, the researcher puts forward the theory that people are becoming more intelligent (as measured by this particular test). She selects a random sample of 500 people all of whom take the test. Their mean score is 103.22.

(ii) Carry out a suitable hypothesis test on the researcher's theory, at the 1% significance level.

SOLUTION

(i) The confidence limits are

$$\bar{x} \pm \frac{k\sigma}{\sqrt{n}} = 105 \pm \frac{k15}{\sqrt{100}}$$

For a two-sided 95% confidence interval, $k = 1.96$.
The confidence interval is $105 - 2.94$ to $105 + 2.94$, i.e. 102.06 to 107.94.

(ii) *Notice that the researcher has correctly collected new data with which to test the hypothesis she formed on the basis of her earlier sample.*

H_0: The parent population mean is unchanged, $\mu = 100$

H_1: The parent population mean has increased, $\mu > 100$

One-tail test at the 1% significance level.

For this sample $n = 500$, $\bar{x} = 103.22$

$$z = \left(\frac{\bar{x} - \mu}{\frac{\sigma}{\sqrt{n}}} \right) = \left(\frac{103.22 - 100}{\frac{15}{\sqrt{500}}} \right) = 4.80.$$

This is to be compared with the critical value. For the 1% significance level

$$z = 2.326.$$

Since 4.80 > 2.236 the null hypothesis is rejected.

The evidence supports the view that scores on this IQ test are now higher; see figure 3.7.

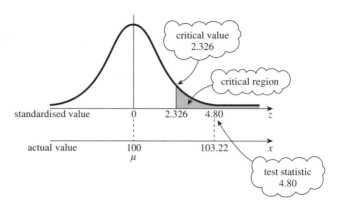

Figure 3.7

 If the test statistic in Example 3.3 had been 2.327, would you have drawn the same conclusion?

---

1 A machine is designed to make paperclips with mean mass 4.00 g and standard deviation 0.08 g. The distribution of the masses of the paperclips is normal. Find

   (i) the probability that an individual paperclip, chosen at random, has mass greater than 4.04 g

   (ii) the standard error of the mass for random samples of 25 paperclips

   (iii) the probability that the mean mass of a random sample of 25 paperclips is greater than 4.04 g.

   A quality control officer weighs a random sample of 25 paperclips and finds their total mass to be 101.2 g.

   (iv) Conduct a hypothesis test at the 5% significance level of whether this provides evidence of an increase in the mean mass of the paperclips. State your null and alternative hypotheses clearly.

2 It is known that the mass of a certain type of lizard has a normal distribution with mean 72.7 g standard deviation 4.8 g. A zoologist finds a colony of lizards in a remote place and is not sure whether they are of the same type. In order to test this, she collects a sample of 12 lizards and weighs them, with the following results:

   80.4   67.2   74.9   78.8   76.5   75.5   80.2   81.9   79.3   70.0   69.2   69.1

   (i) Write down, in precise form, the zoologist's null and alternative hypotheses, and state whether a one-tail or two-tail test is appropriate.

   (ii) Carry out the test at the 5% significance level and write down your conclusion.

   (iii) Would your conclusion have been the same at the 10% significance level?

3 Observations over a long period of time have shown that the mid-day temperature at a particular place during the month of June has a mean value of 23.9 °C with standard deviation 2.3 °C. An ecologist sets up an experiment to collect data for a hypothesis test of whether the climate is getting hotter. She selects at random 20 June days over a five-year period and records the mid-day temperature. Her results (in °C) are as follows:

| 20.1 | 26.2 | 23.3 | 28.9 | 30.4 | 28.4 | 17.3 | 22.7 | 25.1 | 24.2 |
| 15.4 | 26.3 | 19.3 | 24.0 | 19.9 | 30.3 | 32.1 | 26.7 | 27.6 | 23.1 |

   (i) State the null and alternative hypotheses that the ecologist should use.
   (ii) Carry out the test at the 10% significance level and state the conclusion.
   (iii) Calculate the standard deviation of the sample data and comment on it.

4 A zoo has a long established colony of a particular type of rodent which is threatened with extinction in the wild. Observations over several years indicate that the life expectancy for the rodent is 470 days, with standard deviation 45 days. The staff at the zoo suspect that the life expectancy can be increased by improvements to the rodents' environment, and as an experiment allow 36 individuals to spend their whole lives in new surroundings. Their lifetimes are as follows:

491	505	523	470	468	567	512	560	468	498	471	444
511	508	508	421	465	499	486	513	500	488	487	455
523	516	486	478	470	465	487	572	451	513	483	474

   (i) State the null and alternative hypotheses which these data have been collected to test.
   (ii) Carry out the test at the 2% significance level and state the conclusion.
   (iii) How could increased longevity help the rodent population to survive?

5 Some years ago the police did a large survey of the speeds of motorists along a stretch of motorway, timing cars between two bridges. They concluded that their mean speed was 80 mph with standard deviation 10 mph.

Recently the police wanted to investigate whether there had been any change in motorists' mean speed. They timed the first 20 green cars between the same two bridges and calculated their speeds (in mph) to be as follows:

| 85 | 75 | 80 | 102 | 78 | 96 | 124 | 70 | 68 | 92 |
| 84 | 69 | 73 | 78 | 86 | 92 | 108 | 78 | 80 | 84 |

   (i) State suitable null and alternative hypotheses and use the sample data to carry out a hypothesis test at the 5% significance level. State the conclusion.

One of the police officers involved in the investigation says that one of the cars in the sample was being driven exceptionally fast, and that its speed should not be included within the sample data.

   (ii) Would the removal of this outlier alter the conclusion?

6   The keepers of a lighthouse were required to keep records of weather conditions. Analysis of their data from many years showed the visibility at mid-day to have a mean value of 14 sea miles with standard deviation 5.4 sea miles. A new keeper decided he would test his theory that the air had become less clear (and so visibility reduced) by carrying out a hypothesis test on data collected for his first 36 days on duty. His figures (in sea miles) were as follows:

35	21	12	7	2	1.5	1.5	1	0.25	0.25	15	17
18	20	16	11	8	8	9	17	35	35	4	0.25
0.25	5	11	28	35	35	16	2	1	0.5	0.5	1

(i)   Write down suitable null and alternative hypotheses.
(ii) Carry out the test at the 2.5% significance level and state the conclusion that the lighthouse keeper would have come to.
(iii) Criticise the sampling procedure used by the keeper and suggest a better one.

7   The weights of steaks sold by a supermarket are distributed normally with mean μ and standard deviation 0.02 kg. A quality control inspector tests the hypothesis that $\mu = 1$ kg at the 5% level of significance. He takes a random sample of five steaks whose weights (in kg) are:

0.977   1.014   0.989   0.972   0.968

His null hypothesis is that $\mu = 1$ kg, and he performs a two-tailed test. State his alternative hypothesis and perform the test.

Another inspector is employed to check that customers are not (on average) sold underweight steaks. If he had conducted a one-tailed test using the same random sample, the same level of significance and the same null hypothesis, what would have been his alternative hypothesis, and his conclusion?

[MEI]

8   A chemical is packed into bags by a machine. The mean weight of the bags is controlled by the machine operator, but the standard deviation is fixed at 0.96 kg. The mean weight should be 50 kg, but it is suspected that the machine has been set to give underweight bags. If a random sample of 36 bags has a total weight of 1789.20 kg, is there evidence to support the suspicion? (You must state the null and alternative hypotheses and you may assume that the weights of the bags are normally distributed.)

[MEI]

9   The weights of Granny Smith apples are taken as normally distributed with mean 110 grams and standard deviation 8 grams. The apples are sent to the wholesaler in bags which must contain at least 10 kilograms of apples.
(i)   Write down the mean, the standard deviation, and the distribution of the total weight of 90 randomly chosen apples. Show that the probability that this total weight is at least 10 kilograms is approximately 9.4%.

Now suppose that n randomly chosen apples are put into a bag, and that their total weight is T grams.

(ii) Write down the mean and variance of T in terms of n. Show that the condition for $P(T > 10\,000)$ to be at least 99% is

$$10\,000 - 110n \leqslant -18.608\sqrt{n}$$

(iii) Hence find the smallest value of n which gives a probability of at least 99% that the total weight of apples exceeds 10 kilograms.

[MEI]

10 The weights of coffee in tins used by the catering trade are distributed normally with standard deviation 0.071 kg.

A random sample of n tins is taken in order to determine a symmetrical 99% confidence interval for the mean weight μ kg of coffee in a tin. How large should n be for the total width of this interval to be less than 0.05 kg?

In a separate investigation a random sample of 36 tins has a mean weight of 0.981 kg. Test at the 1% significance level the null hypothesis H_0 that $\mu = 1$, the alternative hypothesis H_1 being $\mu < 1$.

Suppose now that it is subsequently discovered that $\mu = 0.9500$. Determine the probability that a significance test at the 1% level, using a new random sample of 36 tins and the same H_0 and H_1 as in the previous paragraph, would give the wrong conclusion, i.e. would lead to acceptance of H_0.

[MEI]

11 Archaeologists have discovered that all skulls found in excavated sites in a certain country belong either to racial group A or to racial group B. The mean lengths of skulls from group A and group B are 190 mm and 196 mm respectively. The standard deviation for each group is 8 mm, and skull lengths are distributed normally and independently.

A new excavation produced 12 skulls of mean length $\bar{x}$ and there is reason to believe that all these skulls belong to group A. It is required to test this belief statistically with the null hypothesis (H_0) that all the skulls belong to group A and the alternative hypothesis (H_1) that all the skulls belong to group B.

(i) State the distribution of the mean length of 12 skulls when H_0 is true.

(ii) Explain why a test of H_0 versus H_1 should take the form:

'Reject H_0 if $\bar{x} > c$',

where c is some critical value.

(iii) Calculate this critical value c to the nearest 0.1 mm when the probability of rejecting H_0 when it is in fact true is chosen to be 0.05.

(iv) Perform the test, given that the lengths (in mm) of the 12 skulls are:

204.1	201.1	187.4	196.4	202.5	185.0
192.6	181.6	194.5	183.2	200.3	202.9

[MEI]

12 A manufacturer of dice makes fair dice and also slightly biased dice which are to be used for demonstration purposes in the teaching of probability and statistics. The probability distributions of the scores for the two types of die are shown in the table.

Score	1	2	3	4	5	6
Probability for fair dice	$\frac{1}{6}$	$\frac{1}{6}$	$\frac{1}{6}$	$\frac{1}{6}$	$\frac{1}{6}$	$\frac{1}{6}$
Probability for biased dice	$\frac{1}{10}$	$\frac{1}{10}$	$\frac{1}{5}$	$\frac{1}{5}$	$\frac{1}{5}$	$\frac{1}{5}$

Calculate the expectation and variance of the score for each type of die.

Unfortunately, some dice have been made without distinguishing marks to show whether they are fair or biased. The manufacturer decides to test such dice as follows: each die is thrown 100 times and the mean score $\bar{x}$ is calculated; if $\bar{x} > 3.7$, the die is classified as biased but, if $\bar{x} \leqslant 3.7$, it is classified as fair. Find the probability that a fair die is wrongly classified as biased as a result of this procedure.

To improve the test procedure the manufacturer increases the number of throws from 100 to N, where N is chosen to make as close as possible to 0.001 the probability of wrongly classifying a biased die as fair. Find the value of N.

[Cambridge]

13 A supermarket's statistician reports that, over the past three months, the mean amount spent per customer has been £43 with a standard deviation of £20.

The supermarket carries out a promotion for one week by offering 'buy two ... get one free' on a range of products which it sells. The management hopes that this will increase the mean amount spent per customer; you may assume that the standard deviation remains unchanged.

A random sample of 50 customers visiting the supermarket that week spent a total of £2400.

(i) Write down suitable null and alternative hypotheses in order to test whether or not the promotion has increased the average level of spending per customer.

(ii) Explain carefully the use of the central limit theorem in carrying out this hypothesis test.

(iii) Carry out the hypothesis test at the 5% significance level, clearly stating your conclusion.

(iv) Find a 90% confidence interval for the mean amount spent by customers during the period of the promotion. State, giving a reason, whether this is consistent with your conclusion in part (iii).

[MEI]

THE AVONFORD STAR

'Honest Gertie' on two charges after fight at Avonford market

Afternoon shoppers found themselves caught up in a fight at Avonford market yesterday. The trouble began when local authority inspector Phil Simmons accused well-known local trader 'Honest Gertie' Haliwell of selling underweight bags of apples.

'They were being sold as 1 kilogram bags', explained Mr Simmons. 'But the six bags I weighed averaged only 955 grams and regulations say they must be 960 or more.'

'It was just the particular bags he chose', said Gertie. 'If he had chosen six others he would most likely have got an average over 1 kilogram. And you've got to think of what each apple weighs.'

'Now Gertie faces two charges, one for assault and one for selling underweight goods. 'I don't know what this country is coming to, treating law abiding citizens like this', Gertie confided to the *Star*.

? Is it likely that Honest Gertie is right and that Phil Simmons just happened to pick an unrepresentative sample?

Incorrect decisions made in hypothesis tests

It is possible to make errors when conducting hypothesis tests. This should not surprise you too much for whenever you make a hypothesis test you are compelled to make a decision about a population parameter on the basis of a single sample of data from a population. The sample, however diligently obtained, may not be a representative one and, even if it is, incorrect decisions can still be made, for example, by relying on confidence intervals to be right all of the time.

Definitions

- A Type I error has occurred if you have rejected the null hypothesis when, in fact, it was true.
- A Type II error has occurred if you have accepted the null hypothesis when, in fact, it was false.

	H_0 **accepted**	H_0 **rejected**
H_0 **true**	Correct result	Type I error
H_0 **false**	Type II error	Correct result

- The probability of making a Type I error is the same as the significance level of the test.

For example, consider the test statistic $z = \dfrac{\bar{x} - \mu}{\frac{\sigma}{\sqrt{n}}}$ used for testing H_0 that the mean of a population is equal to μ.

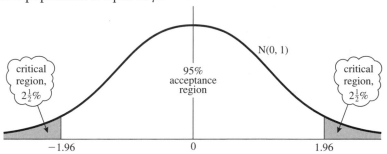

Figure 3.8

Figure 3.8 shows the diagram for a two-tail test at the 5% significance level. The actual measure of the test statistic z will assume a value greater than 1.96 or less than -1.96 on just 5% of occasions when H_0 is true. Thus, if you reject a true null hypothesis, H_0, because $z > 1.96$ or $z < -1.96$, then this is a Type I error and the probability of it occurring is 0.05, the significance level of the test.

However, without further information, you cannot calculate the probability of a Type II error. The fact that H_0 is incorrect leaves you with no particular value of the parameter concerned to use in order to make a probability calculation. If H_0 is an extreme and false contention then the probability of you accepting it and hence making a Type II error will be small, whereas if H_0 seems plausible but is, in fact, false and you incorrectly accept it then this probability may be sizeable.

One measure of the effectiveness of a hypothesis test is to see how good it is at identifying an incorrect null hypothesis. This is called the *power of a test* and is defined, for any given value of the parameter, by:

$$\text{power of test} = 1 - P(\text{Type II error}).$$

Thus $0 \leqslant \text{power} \leqslant 1$ and, other considerations being equal, the larger the power the better the test.

General procedure

At this stage it is instructive to write down the various tasks involved when making a hypothesis test.

- Identify the parent population parameter to be tested.
- Decide on which model is to be used.
- Decide on a significance level; typically this will be 0.1%, 1% or 5%.
- State the null and alternative hypotheses. This will decide whether the test is one- or two-tailed.
- Decide on the test statistic to be used.
- Work out the critical value(s) for the test statistic. You may choose to draw a diagram showing the acceptance and critical region(s).
- Collect the data. Notice that this is not generally the first thing you do!
- Calculate the test statistic. You may wish to indicate it on your diagram.
- If the test statistic falls in the acceptance region then accept the null hypothesis; otherwise reject it.
- Write a conclusion, devoid of technical terms, using phrases like 'the data suggest that ...' or 'the data strongly suggest that ...' but remember that your conclusion is not guaranteed to be the correct one!

EXAMPLE 3.4

A trading standards officer is investigating complaints that a coal merchant is giving short measure. Each sack of coal should contain 25 kg but some variation will inevitably occur because of the size of individual pieces of coal; the officer knows from experience that the standard deviation should be 1.5 kg.

The officer takes a random sample of 36 bags from the merchant's premises in order to investigate the allegation.

(i) Specify an appropriate test procedure given that the officer is working to a 1% significance level. Find the critical region for the test.

(ii) Suppose that the mean mass of a bag is 24.5 kg. What is the probability of a Type II error being made?

(iii) If the mean mass of a bag is 24 kg, what is the probability of a Type II error being made?

SOLUTION

(i) You are testing μ, the mean mass per sack of the merchant's coal. The test is one-tailed at 1% significance level based on the normal distribution.

$$H_0: \mu = 25 \text{ kg}$$
$$H_1: \mu < 25 \text{ kg}.$$

The appropriate model to use is the distribution of the sample mean, $\bar{X}$. Assuming H_0 to be correct, the distribution of $\bar{X}$ is approximately

$N\left(25, \dfrac{2.25}{36}\right)$ and the appropriate test statistic is $z = \dfrac{\bar{x} - 25}{\frac{1.5}{\sqrt{36}}}$

To determine the critical value,

$$\Phi(2.326) = 0.99$$

and so

$$-2.326 = \frac{\bar{x} - 25}{\frac{1.5}{6}}$$

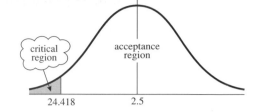

$$\bar{x} = 24.4182 = 24.418 \text{ kg to 3 dp.} \quad \textbf{Figure 3.9}$$

The acceptance region for $\bar{x}$ is $\bar{x} \geqslant 24.418$ kg and the rejection region is $\bar{x} < 24.418$ kg.

(ii) Consider the case $\mu = 24.5$ kg.

The distribution of $\overline{X}$ is approximately $N\left(24.5, \frac{2.25}{36}\right)$. The probability of a Type II error is $P(\overline{X} > 24.4182)$.

Using $z = \dfrac{\bar{x} - \mu}{\frac{\sigma}{\sqrt{n}}}$,

$$z = \frac{24.4182 - 24.5}{\frac{1.5}{6}}$$

$$= -0.3272$$

$$= -0.327 \text{ to 3 dp}$$

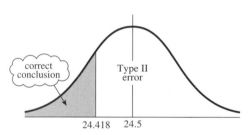

$$\Phi(-0.327) = 0.3719$$

$$P(\text{Type II error}) = 1 - 0.3719$$
$$= 0.6281 \qquad \textbf{Figure 3.10}$$

Notice that the likelihood of making such an error is quite large. This is because the actual mean was only slightly less than that of the null hypothesis. The power of the test when μ is 24.5 is $1 - 0.6281 = 0.3719$, which is low.

(iii) Consider the case $\mu = 24$ kg.

The distribution of $\overline{X}$ is approximately $N\left(24, \frac{2.25}{36}\right)$.

$$P(\text{Type II error}) = P(\overline{X} > 24.4182)$$

Using $z = \dfrac{\bar{x} - \mu}{\frac{\sigma}{\sqrt{n}}}$,

$$z = \frac{24.4182 - 24}{\frac{1.5}{\sqrt{36}}}$$

$$= 1.6728$$

$$= 1.673 \text{ to 3 dp}$$

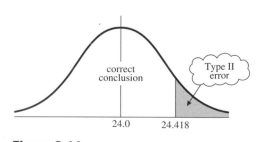

$$\Phi(1.673) = 0.9528$$

$$P(\text{Type II error}) = 1 - 0.9528$$
$$= 0.0472 \qquad \textbf{Figure 3.11}$$

Notice that when the null hypothesis is wrong by a more substantial amount then the probability of a Type II error decreases $(0.0472 \ll 0.6281)$.

The power of the test $= 1 - 0.0472 = 0.9528$. This is much greater than in part (ii).

EXAMPLE 3.5

A die is suspected of bias towards showing more 6s than would be expected of an ordinary die. In order to test this, it is decided to throw the die 12 times. The null hypothesis $p = \frac{1}{6}$, where p is the probability of the die showing a 6, will be rejected in favour of the alternative hypothesis $p > \frac{1}{6}$ if the number of 6s obtained is four or more. Calculate, to three decimal places, the probability of making
(i) a Type I error
(ii) a Type II error if, in fact, $p = \frac{1}{2}$.

Another way of testing the above hypothesis is to throw the die 120 times and reject the null hypothesis if the number of 6s obtained is m or more. Find the value of m for which the probability of making a Type I error is 0.01.

[Cambridge]

SOLUTION

(i) H_0: $p = \frac{1}{6}$

H_1: $p > \frac{1}{6}$

This is a one-tail test, based on the binomial distribution.
Let the random variable X be the number of 6s obtained in 12 throws of the die.

$$\begin{aligned} \text{P(Type I error)} &= \text{P(Rejecting } H_0 \text{ when it is true)} \\ &= \text{P(getting four or more 6s)} \\ &= 1 - \text{P(getting zero, one, two or three 6s)} \\ &= 1 - \left(\tfrac{5}{6}\right)^{12} - 12 \times \left(\tfrac{5}{6}\right)^{11} \times \left(\tfrac{1}{6}\right)^{1} \\ &\quad - {}^{12}C_2 \times \left(\tfrac{5}{6}\right)^{10} \times \left(\tfrac{1}{6}\right)^{2} - {}^{12}C_3 \times \left(\tfrac{5}{6}\right)^{9} \times \left(\tfrac{1}{6}\right)^{3} \\ &= 0.125 \text{ to 3 dp} \end{aligned}$$

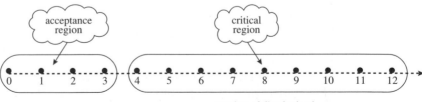

number of 6's obtained

Figure 3.12

(ii) $P(\text{Type II error} \mid p = \frac{1}{2}) = P(\text{accepting } H_0 \mid p = \frac{1}{2})$

$$= P(X < 4 \mid p = \frac{1}{2})$$

$$= \left(\frac{1}{2}\right)^{12} + 12 \times \left(\frac{1}{2}\right)^{11} \times \left(\frac{1}{2}\right)^1 + {}^{12}C_2 \times \left(\frac{1}{2}\right)^{10} \times \left(\frac{1}{2}\right)^2$$

$$+ {}^{12}C_3 \times \left(\frac{1}{2}\right)^9 \times \left(\frac{1}{2}\right)^3$$

$$= 0.073 \text{ to 3 dp}$$

If the die is rolled 120 times, let the random variable Y denote the number of 6s obtained.

Y has a binomial distribution with $n = 120$, $p = \frac{1}{6}$ and $q = \frac{5}{6}$.

Since n is large, $Y \sim N(120 \times \frac{1}{6}, 120 \times \frac{1}{6} \times \frac{5}{6})$

$$Y \sim N(20, 16.\dot{6})$$

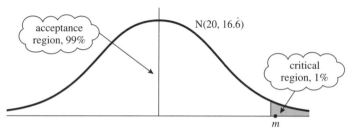

Figure 3.13

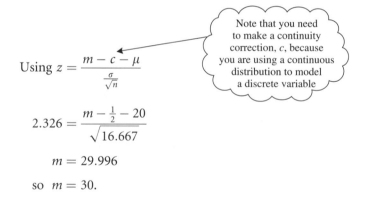

Using $z = \dfrac{m - c - \mu}{\frac{\sigma}{\sqrt{n}}}$

> Note that you need to make a continuity correction, c, because you are using a continuous distribution to model a discrete variable

$$2.326 = \frac{m - \frac{1}{2} - 20}{\sqrt{16.667}}$$

$$m = 29.996$$

so $m = 30$.

1 The null hypothesis, $\mu = 10$, where μ is the population mean, is tested against the alternative hypothesis $\mu \neq 10$ at the 1% significance level. What is the probability of making a Type I error?

If a Type I error is made,

(i) state the conclusion of the test

(ii) state which hypothesis is false.

[Cambridge]

2 A random variable has a normal distribution with mean μ and standard deviation 8. The null hypothesis $\mu = 12$ is to be tested against the alternative hypothesis $\mu > 12$ using a random sample of size 16. The null hypothesis will be rejected if the sample mean is greater than 14.7.

Calculate the probability of making a Type I error.

[Cambridge]

3　The playing times of a particular brand of audio tape are normally distributed with mean μ minutes and standard deviation 0.24 minutes. The manufacturer states that $\mu = 60$. A large batch of these tapes is delivered to a store and, in order to check the manufacturer's statement, the playing times of a random sample of ten tapes were measured. The null hypothesis $\mu = 60$ is tested against the alternative hypothesis $\mu < 60$ at the 1% significance level.

(i)　Find the range of values of the sample mean $\overline{X}$ for which the null hypothesis is rejected, giving 2 decimal places in your answer.

(ii)　Calculate the probability of making a Type II error when, in fact, $\mu = 59.7$.

[Cambridge]

4　The breaking strength of a certain type of fishing line has a normal distribution with standard deviation 0.24 kN. A random sample of ten lines is tested. The mean breaking strengths of the sample and of the population are $\overline{x}$ kN and μ kN respectively. The null hypothesis $\mu = 8.75$ is tested against the alternative hypothesis $\mu < 8.75$ at the $2\frac{1}{2}$% significance level.

(i)　Show that the range of values of $\overline{x}$ for which the null hypothesis is rejected is given by $\overline{x} < 8.60$, correct to 2 decimal places.

(ii)　State the probability of making a Type I error.

(iii)　Find the probability of making a Type II error when $\mu = 8.50$.

(iv)　State the power of the test when $\mu = 8.50$.

[Cambridge]

5　Boxes of a certain breakfast cereal have contents whose masses are normally distributed with mean μ grams and standard deviation 15 grams. A test of the null hypothesis $\mu = 375$ against the alternative hypothesis $\mu > 375$ is carried out at the $2\frac{1}{2}$% significance level using a random sample of 16 boxes.

(i)　Show that the alternative hypothesis is accepted when $\overline{x} > 382.35$, where $\overline{x}$ grams is the sample mean mass.

(ii)　Given that the actual value of μ is 385, find the probability of making a Type II error.

(iii)　State the power of the test with this value of μ.

[Cambridge]

6　When cars arrive at a certain T-junction they turn either right or left. Part of a study of road usage involved deciding between the following alternatives:

cars are equally likely to turn right or left
cars are more likely to turn right than left.

(i)　State suitable null and alternative hypotheses, involving a probability, for a significance test.

(ii)　Out of a random sample of 40 cars, n turned right. Use a suitable approximation to find the least value of n for which the null hypothesis will be rejected at the 2% significance level.

(iii)　For the test described in part (ii), calculate the probability of making a Type II error when, in fact, 80% of all cars arriving at the junction turn right.

[Cambridge]

For situations where the population mean, μ, is unknown but the population variance, σ^2 (or standard deviation, σ), is known:

The central limit theorem

1   For samples of size n drawn from a distribution with mean μ and finite variance σ^2, the distribution of the sample mean is approximately $N\left(\mu, \dfrac{\sigma^2}{n}\right)$ for sufficiently large n.

The standard error of the mean

2   The standard error of the mean (i.e. the standard deviation of the sample means) is given by $\dfrac{\sigma}{\sqrt{n}}$.

Confidence intervals

3   Two-sided confidence intervals for μ are given by

$$\bar{x} - \frac{k\sigma}{\sqrt{n}} \quad \text{to} \quad \bar{x} + \frac{k\sigma}{\sqrt{n}}$$

4   The value of k for any confidence level can be found using normal distribution tables

Confidence level	k
90%	1.645
95%	1.96
99%	2.58

Hypothesis testing

5   Sample data may be used to carry out a hypothesis test on the null hypothesis that the population mean has some particular value, μ.

6   The test statistic is $z = \left(\dfrac{\bar{x} - \mu}{\frac{\sigma}{\sqrt{n}}}\right)$ and the normal distribution is used.

7   A Type I error has occurred if you have rejected the null hypothesis when, in fact, it was true.

8   A Type II error has occurred if you have accepted the null hypothesis when, in fact, it was false.

	H_0 **accepted**	H_0 **rejected**
H_0 **true**	Correct result	Type I error
H_0 **false**	Type II error	Correct result

9   The probability of making a Type I error is the same as the significance level of the test.

10   The power of a test $= 1 - P(\text{Type II error})$.

4

Interpreting sample data using the *t*-distribution

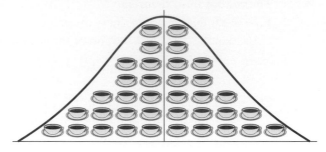

The tea distribution was quite normal.

Douglas Butler: MEI Newsletter report on
an annual general meeting

THE AVONFORD STAR

Local students find new bat

From our Science Correspondent Ama Williams

Two students and a lecturer from Avonford
Community College have found their way into
the textbooks. On a recent field trip they
discovered a small colony of a previously
unknown bat living in a cave.

'Somewhere in Britain' is all that Shakila
Mahadavan, 20, would tell me about its location.
'We don't want the general public disturbing the
bats or worse still catching them for
specimens', she explained.

The other two members of the group, lecturer
Alison Evans and 21-year-old Iain Scott, showed
me scores of photographs of the bats as well as
pages of measurements that they had gently
made on the few they had caught before
releasing them back into their cave.

*At a mystery location, Avonford students have
pushed forward the frontiers of science*

The measurements referred to in the article include the weights (in g) of eight
bats which were identified as adult males:

$$156 \quad 132 \quad 160 \quad 142 \quad 145 \quad 138 \quad 151 \quad 144$$

From these figures, the team want to estimate the mean weight of an adult male
bat, and 95% confidence limits for their figure.

It is clear from the newspaper report that these are the only measurements
available. All that is known about the parent population is what can be inferred
from these eight measurements. You know neither the mean nor the standard
deviation of the parent population, but you can estimate both.

The mean is estimated to be the same as the sample mean:

$$\frac{156 + 132 + 160 + 142 + 145 + 138 + 151 + 144}{8} = 146.$$

When it comes to estimating the standard deviation you have to be careful. The deviations of the eight numbers are:

$$156 - 146 = \quad 10$$
$$132 - 146 = -14$$
$$160 - 146 = \quad 14$$
$$142 - 146 = \quad -4$$
$$145 - 146 = \quad -1$$
$$138 - 146 = \quad -8$$
$$151 - 146 = \quad +5$$
$$144 - 146 = \quad -2$$

These eight deviations are not independent: they must add up to zero because of the way the mean is calculated. This means that when you have worked out the first seven deviations, it is inevitable that the final one has the value it does (in this case -2). Only seven values of the deviation are independent, and in general only $n - 1$ out of the n deviations from the sample mean are independent.

Consequently the variance is worked out using the formula:

$$\sum_i \frac{(x_i - \bar{x})^2}{(n - 1)}$$

When the variance is worked out in this way, the resulting value is an *unbiased estimate of the parent population variance*. A particular value is denoted by s^2, the associated random variable by S^2.

$$s^2 = \sum_i \frac{(x_i - \bar{x})^2}{(n - 1)}$$

In the case of the bats this gives

$$s^2 = \frac{(100 + 196 + 196 + 16 + 1 + 64 + 25 + 4)}{7} = 86.$$

The corresponding value of the standard deviation is $s = 9.27$.

Degrees of freedom

The use of $n - 1$ in the calculation of s^2 illustrates an important idea in statistics. Much of the theory of statistics involves considering a particular sample in relation to all possible samples of the same size, subject to any restrictions. In this case there is one restriction, the mean. For a given value of the mean, $n - 1$ of the data items are free to take any reasonable value, but the value of the last one is then fixed if the mean is to work out correctly.

Consequently there are $n - 1$ *free variables* in this situation. The number of free variables within a system is called the *degrees of freedom* and denoted by v.

You need to know the degrees of freedom in many situations where you are calculating confidence intervals or conducting hypothesis tests. You may recall meeting the idea in Chapter 4 of *Statistics 2* when interpreting correlation coefficients.

Calculating the confidence intervals

Returning to the problem of estimating the mean weight of the bats, you now know that:

$$\bar{x} = 146, \quad s^2 = 86, \quad s = 9.27, \quad \text{and} \quad v = 8 - 1 = 7.$$

Before starting on further calculations there are some important and related points to notice.

1 This is a small sample. It would have been much better if they had managed to catch and weigh more than eight bats.

2 The true parent standard deviation, σ, is unknown and consequently the standard deviation of the sampling distribution given by the central limit theorem, $\dfrac{\sigma}{\sqrt{n}}$, is also unknown.

3 In situations where the sample is small and the parent standard deviation or variance is unknown, there is little more that can be done unless you can assume that the parent population is normal. (In this case that is a reasonable assumption, the bats being a naturally occurring population.) If you can assume normality, then you may use the *t*-distribution, estimating the value of σ from your sample.

4 It is possible to test whether a set of data could reasonably have been taken from a normal distribution by using normal probability paper. The method involves making a cumulative frequency table and plotting points on a graph with specially chosen axes. If the graph obtained is approximately a straight line then the data could plausibly have been drawn from a normal population. Otherwise a normal population is unlikely.

The *t*-distribution looks very like the normal distribution, and indeed for large values of v is little different from it. The larger the value of v, the closer the *t*-distribution is to the normal. Figure 4.1 shows the normal distribution and *t*-distributions $v = 2$ and $v = 10$.

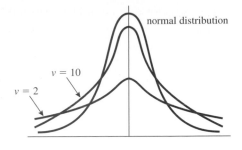

normal distribution

$v = 10$

$v = 2$

Figure 4.1

Historical note

William S. Gosset was born in Canterbury in 1876. After studying both mathematics and chemistry at Oxford he joined the Guinness breweries in Dublin as a scientist. He found an immense amount of statistical data was available, relating the brewing methods and the quality of the ingredients, particularly barley and hops, to the finished product. Much of these data were in the form of samples, and Gosset developed techniques to handle it, including the discovery of the *t*-distribution. Gosset published his work under the pseudonym 'Student' and so the t-test is often called Student's t-test.

Gosset's name has frequently been misspelt as Gossett (with a double t), giving rise to puns about the *t*-distribution.

Confidence intervals using the *t*-distribution are constructed in much the same way as those using the normal, with the confidence limits given by:

$$\bar{x} \pm \frac{ks}{\sqrt{n}}$$

where the values of k are found from a table of percentage points of the *t*-distribution, for the appropriate degrees of freedom, v, and confidence level.

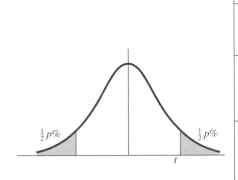

$\frac{1}{2}p\%$ $\frac{1}{2}p\%$

t

$p\%$	10	⑤	2	1	
$v = 1$	6.314	12.71	31.82	63.66	
2	2.920	4.303	6.965	9.925	
3	2.353	3.182	4.541	5.841	
4	2.132	2.776	3.747	4.604	
5	2.015	2.571	3.365	4.032	
6	1.943	2.447	3.143	3.707	
⑦	1.895	2.365	2.998	3.499	$v = 7, p = 5\%$
8	1.860	2.306	2.896	3.355	gives $k = 2.365$
9	1.833	2.262	2.821	3.250	
10	1.812	2.228	2.764	3.169	
11	1.796	2.201	2.718	3.106	
12	1.782	2.179	2.681	3.055	
13	1.771	2.160	2.650	3.012	
14	1.761	2.145	2.624	2.977	
15	1.753	2.131	2.602	2.947	
20	1.725	2.086	2.528	2.845	
30	1.697	2.042	2.457	2.750	
50	1.676	2.009	2.403	2.678	
100	1.660	1.984	2.364	2.626	
∞	1.645	1.960	2.326	2.576	= Percentage points of the normal distribution N(0, 1)

Figure 4.2

To construct a 95% confidence interval for the mean weight of the bats, you look under $p = 5\%$ and $v = 7$, to get $k = 2.365$; see figure 4.2. This gives a 95% confidence interval of

$$146 - \frac{2.365 \times 9.27}{\sqrt{8}} \quad \text{to} \quad 146 + \frac{2.365 \times 9.27}{\sqrt{8}}$$

$$138.2 \quad \text{to} \quad 153.8$$

Hypothesis testing on a sample mean using the *t*-distribution

THE AVONFORD STAR

Letters to the editor

Dear Sir,
It was with great interest that I read about the discovery of a previously unknown type of bat. My grandfather was a keen naturalist and I well-remember him telling us that he had found a cave containing a large colony of bats not known to science. He refused to tell anybody where the cave was but he spent months there observing and measuring them.

Unfortunately he perished when his house was hit by a stray bomb during the war, and all his notes went up in flames. I do have an old diary of his which includes a scribbled note 'Average weight of male bats 160 g, females 154.9 g'.

Could these be the same bats?

Yours faithfully,

Julia Bainton.

It will never be possible to know for certain whether the two colonies of bats are of the same type. It is even conceivable that Julia's grandfather had found the same cave and that the students had discovered the descendants of his bats, but the information has been irretrievably lost.

What you can do is to set up a hypothesis test that the newly-discovered bats are drawn from a population with the same mean weight for adult males of 160 g.

H_0: $\mu = 160$

H_0: $\mu \neq 160$

Two-tail test, 1% significance level.

The sample data may be summarised (figure 4.3) by:

$$\bar{x} = 146, \quad s = 9.27, \quad \text{and} \quad v = 8 - 1 = 7.$$

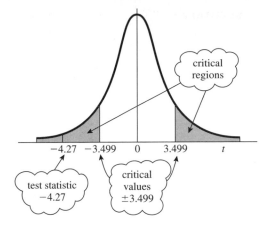

Figure 4.3

The test statistic is $\quad t = \dfrac{\bar{x} - \mu}{\frac{s}{\sqrt{n}}}$

and this has value $\quad t = \dfrac{146 - 160}{\frac{9.27}{\sqrt{8}}} = -4.27.$

This is compared with the critical value 3.499 found in the *t*-distribution tables under $p = 1\%$ and $v = 7$.

Since $-4.27 < -3.499$ the null hypothesis is rejected. The evidence suggests the bats were not of the same type.

Note

Underlying this test is the assumption that the distribution of the weights of the students' bats is normal. If this is not the case, the test is not valid.

EXAMPLE 4.1

Tests are being carried out on a new sleeping pill which is given one evening to a random sample of 16 people. The number of hours they sleep may be assumed to be normally distributed and is recorded as follows:

$$8.1 \quad 6.7 \quad 3.3 \quad 7.2 \quad 8.1 \quad 9.2 \quad 6.0 \quad 7.4$$
$$6.4 \quad 6.9 \quad 7.0 \quad 7.8 \quad 6.7 \quad 7.2 \quad 7.6 \quad 7.9$$

(i) Use these data to set up a 95% confidence interval for the mean length of time somebody sleeps after taking the pill.

The mean number of hours slept by a large control group is 6.6.

(ii) Carry out a test, at the 1% significance level, of the hypothesis that the new drug increases the number of hours a patient sleeps.

SOLUTION

(i) For the given data,

$$n = 16, \quad v = 16 - 1 = 15, \quad \bar{x} = 7.093\,75, \quad s = 1.276.$$

For a 95% confidence interval, with $v = 15$, $k = 2.131$ (from tables).

The confidence limits are given by

$$\bar{x} \pm \frac{ks}{\sqrt{n}} = 7.093\,75 \pm \frac{2.131 \times 1.276}{\sqrt{16}}$$

So the 95% confidence interval for μ is 6.41 to 7.77.

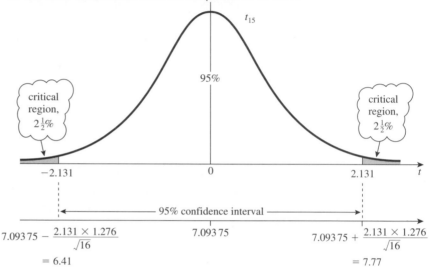

Figure 4.4

(ii) H_0: There is no change in the mean number of hours sleep. $\mu = 6.6$

H_1: There is an increase in the mean number of hours sleep. $\mu > 6.6$

One-tail test at the 1% significance level.

For this sample

$$n = 16, \quad v = 16 - 1 = 15, \quad \bar{x} = 7.094, \quad s = 1.276.$$

The critical value for t, for $v = 15$, at the 1% significance level, is found from tables to be 2.602

$$t = \left(\frac{\bar{x} - \mu}{\frac{s}{\sqrt{n}}} \right) = \left(\frac{7.094 - 6.6}{\frac{1.276}{\sqrt{16}}} \right) = 1.55$$

This is to be compared with 2.602, the critical value, for the 1% significance level.

Since $1.55 < 2.602$, there is no reason at the 1% significance level to reject the null hypothesis that the mean number of hours sleep remains the same.

1   An aptitude test for entrance to university is designed to produce scores which may be modelled by the normal distribution. In early testing 15 students from the appropriate age group are given the test. Their scores (out of 500) are as follows:

321   445   219   378   317   407   289   345

276   463   265   165   340   298   315

(i) Use these data to estimate the mean and standard deviation to be expected for students taking this test.

(ii) Construct a 95% confidence interval for the mean.

2   A fruit farmer has a large number of almond trees, all of the same variety and of the same age. One year he wishes to estimate the mean yield of his trees. He collects all the almonds from eight trees and records the following weights (in kg):

36   53   78   67   92   77   59   66

(i)   Use these data to estimate the mean and standard deviation of the yields of all the farmer's trees.

(ii) Construct a 95% confidence interval for the mean yield.

(iii) What statistical assumption is required for your procedure to be valid?

(iv) How might you select a sample of eight trees from those growing in a large field?

3   A fair trading inspector visits a butcher who sells meat pies. The inspector investigates the meat content of 12 pies and records these figures (in grams):

234   256   171   234   251   251   243   216   251   232   250   253

(i)   Use these figures to construct a 90% confidence interval for the mean mass of meat in one of the butcher's pies.

The pies are supposed to contain at least 250 g of meat each but there have been complaints that the butcher does not put enough meat in (which is why the inspector went to this shop). These data are used to carry out a suitable hypothesis test.

(ii) State the null and alternative hypotheses.

(iii) Carry out the test at the 5% significance level. State any assumptions you make and your conclusion.

The butcher says 'I don't know what all the fuss is about. Some are bound to be a bit over and some a bit under but you can count for yourself. Half the ones the inspector took are at least the right weight.'

(iv) Comment.

4 A forensic scientist is trying to decide whether a man accused of fraud could have written a particular letter. As part of the investigation she looks at the lengths of sentences used in the letter. She finds them to have the following numbers of words:

$$17 \quad 18 \quad 25 \quad 14 \quad 18 \quad 16 \quad 14 \quad 16 \quad 16 \quad 21 \quad 25 \quad 19$$

(i) Use these data to estimate the mean and standard deviation of the lengths of sentences used by the letter writer.

(ii) Construct a 90% confidence interval for the mean length of the letter writer's sentences.

(iii) What assumptions have you made to obtain your answer?

While the accused man is awaiting trial he writes a very large number of letters protesting his innocence. Many of these are passed on to the forensic scientist who is able to establish that overall the man uses a mean of 15.5 words per sentence. The scientist then uses the data from the original letter to carry out a suitable hypothesis test at the 5% significance level.

(iv) State the null and alternative hypotheses for the scientist's test.

(v) Carry out the test and state the conclusion.

(vi) Does this evidence support a prosecution?

5 A large company is investigating the number of incoming telephone calls at its exchange, in order to determine how many telephone lines it should have. During March one year the number of calls received each day was recorded and written down, across the page, as follows:

623	584	598	701	656	210	23	655	661	599
634	681	197	25	592	643	642	698	659	201
19	588	672	612	706	650	212	29	681	642
677

(i) What day of the week was 1 March?

(ii) Which of the data do you consider relevant to the company's research and why?

(iii) Construct a 95% confidence interval for the number of incoming calls per weekday.

(iv) Your calculation is criticised on the grounds that your data are discrete and so the underlying distribution cannot possibly be normal. How would you respond to this criticism?

6 A tyre company is trying out a new tread pattern which it is hoped will result in the tyres giving greater distance. In a pilot experiment, 12 tyres are tested; the mileages ($\times$ 1000 miles) at which they are condemned are as follows:

$$65 \quad 63 \quad 71 \quad 78 \quad 65 \quad 69 \quad 59 \quad 81 \quad 72 \quad 66 \quad 63 \quad 62$$

(i) Construct a 95% confidence interval for the mean distance that a tyre travels before being condemned.

(ii) What assumptions, statistical and practical, are required for your answer to part (i) to be valid?

Tyres with the company's usual tread pattern have been found to travel a mean distance of 62 000 miles before being condemned.

(iii) Carry out a test at the 0.5% significance level to determine whether the new tread gives a greater distance. State clearly your null and alternative hypotheses and your conclusion.

7 A fisherman claims that pollack are not as big as they used to be. 'They used to average three quarters of a kilogram each', he says. When challenged to prove his point he catches 20 pollack from the same shoal. Their masses (in kg) are as follows:

$$0.65 \quad 0.68 \quad 0.77 \quad 0.71 \quad 0.67 \quad 0.75 \quad 0.69 \quad 0.72 \quad 0.73 \quad 0.69$$
$$0.70 \quad 0.70 \quad 0.72 \quad 0.76 \quad 0.73 \quad 0.78 \quad 0.75 \quad 0.69 \quad 0.70 \quad 0.71$$

(i) State the null and alternative hypotheses for this test.

(ii) Carry out the test at the 5% significance level and state the conclusion.

(iii) State any assumptions underlying your procedure and comment on their validity.

8 In the game of bridge a standard pack of 52 playing cards is dealt into four hands of 13 cards each. Players usually assess the value of their hands by counting 4 points for an ace, 3 for a King, 2 for a Queen, 1 for a Jack and nothing for any other card. The total points available from the four suits are $(4 + 3 + 2 + 1) \times 4 = 40$. So the mean number of points per hand is $\frac{40}{4} = 10$. Helene claims that she never gets good cards. One day she is challenged to prove this and agrees to keep a record of the number of points she gets on each hand next time she plays, with the following results:

$$5 \quad 16 \quad 7 \quad 1 \quad 11 \quad 2 \quad 8 \quad 9 \quad 14 \quad 12$$
$$21 \quad 10 \quad 0 \quad 7 \quad 12 \quad 7 \quad 6 \quad 8 \quad 13 \quad 4$$

(i) What assumption underlies the use of the t-test in this situation? To what extent do you think the assumption is justified?

(ii) State null and alternative hypotheses relating to Helene's claim.

(iii) Carry out the test at the 5% significance level and comment on the result.

(This question is set in memory of a lady called Helene who claimed that bridge hands had not been the same since the Second World War.)

9 A bus company is about to start a scheduled service between two towns some distance apart. Before deciding on an appropriate timetable they do ten trial runs to see how long the journey takes. The times, in minutes, are:

$$89 \quad 92 \quad 95 \quad 94 \quad 88 \quad 152 \quad 90 \quad 92 \quad 93 \quad 91$$

(i) Construct a 95% confidence interval for the mean journey time, justifying any decision you make with regard to the data.

(ii) Explain why the company might actually be more interested in a one-sided confidence interval than in the two-sided one you have just calculated.

The company regards its main competition as the railway service, which takes 95 minutes, and claims that the bus journey time is less.

(iii) Use the sample data to test at the 5% significance level whether the company's claim is justified.

10 A history student wishes to estimate the life expectancy of people in Lincolnshire villages around 1750. She looks at the parish registers for five villages at that time and writes down the ages of the first ten people buried after the start of 1750. Those less than one year old were recorded as 0. The data were as follows:

2	6	72	0	0	18	45	91	6	2
0	12	56	4	25	1	1	5	0	7
8	65	12	63	2	76	70	0	1	0
9	15	3	49	54	0	2	71	6	8
6	0	67	55	2	0	1	54	1	5

(i) Use these data to estimate the mean life expectancy at that time.

(ii) Explain why it is not possible to use these data to construct a confidence interval for the mean life expectancy.

(iii) Is a confidence interval a useful measure in this situation anyway?

A friend tells the student that she could construct a confidence interval for the mean life expectancy of those who survive childhood (age $\geqslant$ 15).

(iv) Construct a 95% confidence interval for the mean life expectancy of this group, and comment on whether you think your procedure is valid.

11 A large fishing-boat made a catch of 500 mackerel from a shoal. The total mass of the catch was 320 kg. The standard deviation of the mass of individual mackerel is known to be 0.06 kg.

Find a 99% confidence interval for the mean mass of a mackerel in the shoal.

An individual fisherman caught ten mackerel from the same shoal. These had masses (in kg) of

1.04   0.94   0.92   0.85   0.85   0.70   0.68   0.62   0.61   0.59

(i) From these data only, use your calculator to estimate the mean and standard deviation of the masses of mackerel in the shoal.

(ii) If the masses of mackerel are assumed to be normally distributed, use your results from (i) to find another 99% confidence interval for the mean mass of a mackerel in the shoal.

(iii) Give two statistical reasons why you would use the first limits you calculated in preference to the second limits.

12 A random sample of ten independent observations of a normally distributed random variable X is taken from a population, and a test statistic, $t = 3.1$, is calculated. It is thought that the population mean μ is 110.0. Write down a suitable null hypothesis and a suitable alternative hypothesis for a two-tail significance test for μ. Use a t-test to test your null hypothesis against your alternative hypothesis at the 1% significance level.

[Cambridge]

13 In a classroom experiment to estimate the mean height, μ cm, of seventeen-year-old boys, the heights, x cm, of ten such pupils were obtained. The data were summarised by

$$\Sigma x = 1727, \quad \Sigma x^2 = 298\,834.$$

(i) Find the mean and variance of the data, and use them to find the symmetrical 95% confidence interval of μ. State clearly but briefly the two important assumptions which you need to make.

An experiment is planned using the heights of 150 seventeen-year-old boys.

(ii) What effect will the use of a larger sample have on the width of the confidence interval for μ? Identify two distinct mathematical reasons for this effect.

(iii) To what extent are the assumptions made in part (i) still necessary with the larger sample size?

[MEI]

14 A rail commuter suspects that the train he travels to work by arrives late regularly. On five such journeys he noted the times, T minutes, by which the train was late. The data were summarised by

$$\Sigma t = 11.8, \quad \Sigma t^2 = 65.3.$$

(i) Find the mean and standard deviation of the data, and use them to find a and b, the lower and upper limits for the 90% confidence interval for μ, the mean time by which the train is late. State clearly two important assumptions you need to make.

(ii) The commuter interprets the result in (i) as 'the probability that μ lies between a and b is 0.9'. Comment on this interpretation.

(iii) Use your result in part (i) to test whether the commuter's suspicions are justified. State the significance level at which the test is carried out.

(iv) The railway management notes the value of T on a further 40 occasions. State, with reasons, how you might expect its confidence interval for μ to differ from the one calculated in part (i).

[MEI]

15 When a darts player aims at the centre of the dart board the distance from the centre to the point where the dart lands is R, which is modelled as a continuous random variable. A suggested probability density function for R is as shown in the diagram.

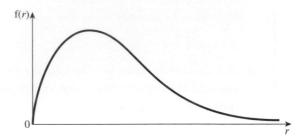

(i) Copy the diagram and show the position of the mode. Show also the approximate position of the mean in relation to the mode.

The mean value of R will vary for players of different ability. The famous darts player Willy Jackson aims 50 darts at the centre of the board. The distances from the centre are summarised as follows.

$$\Sigma r = 35.5, \quad \Sigma r^2 = 33.2$$

(ii) Construct a 90% confidence interval for the mean value of R. What assumption(s) have you made about the set of 50 throws?

(iii) Estimate the number of throws required if the 90% confidence interval for the mean of R for Willy Jackson is to be of width 0.1. Explain why your answer is only an estimate.

(iv) A second darts player records the results for 10 throws. Explain carefully why it would not be possible to obtain an accurate confidence interval for the mean value of R for this player.

[MEI]

16 A Youth Club has a large number of members (referred to as the *population* in the remainder of the question). In order to find the distribution of weekly allowances of the members, a random sample of ten is questioned.

(i) Describe a method of producing the random sample.

Such a random sample produced the following weekly allowances:

£5.20, £4.40, £3.00, £2.00, £3.30,

£7.50, £5.00, £6.50, £4.80, £5.70.

(ii) Estimate the population mean and variance.

(iii) Find a 95% confidence interval for the population mean. State any assumptions on which your method is based.

(iv) Explain how the width of the confidence interval may be reduced. Assuming the same variance as in part (ii) what must the sample size be to reduce the width to £2?

[MEI]

17 When watching games of men's basketball, I have noticed that the players are often tall. I am interested to find out whether or not men who play basketball really are taller than men in general.

I know that the heights, in metres, of men in general have the distribution $N(1.73, 0.08^2)$. I make the assumption that the heights, X metres, of male basketball players are also normally distributed, with the same variance as the heights of men in general, but possibly with a larger mean.

(i) Write down the null and alternative hypotheses under test.

I propose to base my test on the heights of eight male basketball players who recently appeared for our local team, and I shall use a 5% level of significance.

(ii) Write down the distribution of the sample mean, $\overline{X}$, for samples of size eight drawn from the distribution of X, assuming that the null hypothesis is true.

(iii) Determine the critical region for my test, illustrating your answer with a sketch.

(iv) Carry out the test, given that the mean height of the eight players is 1.765 m. You should present your conclusions carefully, stating any additional assumption you need to make.

In fact, the distribution of X is $N(1.80, 0.06^2)$.

(v) Find the probability that a test based on a random sample of size eight and using the critical region in part (iii) will lead to the conclusion that male basketball players are *not* taller than men in general.

[MEI]

18 An experiment to determine the acceleration due to gravity, g ms^{-2}, involves measuring the time, T seconds, taken by a pendulum of length 1 m to perform complete swings. T is regarded as a random variable.

30 measurements are made on T, and they are summarised by

$$\Sigma t = 59.8, \quad \Sigma t^2 = 119.7.$$

Construct a two-sided 98% confidence interval for μ, the mean value of T. Determine the corresponding range of values of g, using the formula

$$g = \frac{4\pi^2}{\mu^2}.$$

This result for g is not precise enough, so a longer series of measurements of T is made. Assuming that the sample mean and standard deviation remain about the same, how many measurements will be required in total to halve the width of the 98% confidence interval for μ? What will be the corresponding effect on the range of values for g?

[MEI]

1 For situations where the population mean, μ, and variance, σ^2 (or standard deviation, σ), are both unknown, sample data may be interpreted using the *t*-distribution provided the distribution from which they are drawn is normal.

Confidence intervals

2 Two-sided confidence intervals for μ are given by $\bar{x} - \dfrac{k\sigma}{\sqrt{n}}$ to $\bar{x} + \dfrac{k\sigma}{\sqrt{n}}$.

3 The degrees of freedom, v, are given by $v = n - 1$.

4 The value of k for any confidence level can be found using tables of critical values (or percentage points) for the *t*-distribution.

Hypothesis testing

5 Sample data may be used to carry out a hypothesis test on the null hypothesis that the population mean has some particular value, μ.

The test statistic is $t = \left(\dfrac{\bar{x} - \mu}{\frac{s}{\sqrt{n}}}\right)$ and the *t*-distribution is used.

6 **The distribution of the mean, $\bar{x}$, of a sample**

Population: mean μ, variance σ^2

Sample: $x_1, x_2, \ldots, x_n$

$$\bar{x} = \frac{(x_1 + x_2 + \ldots + x_n)}{n} \qquad s^2 = \frac{\sum (x_i - \bar{x})^2}{(n-1)}$$

Population variance	Underlying population	Sampling distribution	
Known The variance is σ.	Any population.	The distribution is $N\left(\mu, \dfrac{\sigma^2}{n}\right)$ for all samples.	
Unknown The variance is estimated by s^2.	Normal	The distribution variance $\dfrac{s^2}{n}$ and freedom for all is *t*-, with mean μ, and $n-1$ degrees of samples.	
		Large samples	**Small samples**
		You may also use $N\left(\mu, \dfrac{s^2}{n}\right)$.	You must use the *t*-distribution.
	Not normal	You may use the distribution $N\left(\mu, \dfrac{s^2}{n}\right)$	Nothing useful can be done.

5 The χ^2 (chi squared) distribution

The fact that the criterion which we happen to use has a fine ancestry of statistical theorems does not justify its use. Such justification must come from empirical evidence that it works.

W. A. Shewhart

THE AVONFORD STAR

The die that Roald rolled

Those of you who watched the big match last night will have missed Jane McNulty's fascinating documentary 'Mind Over Matter?' In her usual incisive and questioning style Ms McNulty explored the extraordinary world of Roald Drysdale.

Those of us who live in Avonford have long known of Roald and his equally extraordinary mother Blanche. Roald's claims to be able to influence the world around him by mind alone are just as remarkable as Uri Geller's spoon bending.

I was fascinated by his demonstration of his influence over a die being thrown. 'You must realise that objects we take to be inanimate do in fact have a spirit,' he explained. 'Take this die. At the moment its spirit is willing it to produce a certain result. I can't change that, but by concentrating my Thought Field on the die, I can enhance its spirit and help it to produce the outcome that it is seeking.'

The power of the mind – Roald Drysdale claims to have psychic influence

That evening the die's spirit was clearly willing it to show 1, as you can see from these remarkable results from 120 throws.

Score	6	5	4	3	2	1
Frequency	12	16	15	23	24	30

Do these figures provide evidence that Roald had influenced the die, or is this just the level of variation you would expect to occur naturally? Clearly a formal statistical test is required.

If Roald's claim had been that he could make a particular number, say 6, turn up more often than the other numbers, you could use a binomial test.

However, all he said was that he could 'enhance the spirit of the die' by making some number come up more than the others. So all six outcomes are involved and a different test is needed.

The expected distribution of the results, based on the null hypothesis that the die is not biased, is easily obtained. The probability of each outcome is $\frac{1}{6}$ and so the expectation for each number is $120 \times \frac{1}{6} = 20$.

Outcome	1	2	3	4	5	6
Expected frequency, f_e	20	20	20	20	20	20

You would not, however, expect exactly this result from 120 throws. Indeed you would be very suspicious if somebody claimed to have obtained it, and might well disbelieve it. You expect random variation to produce small differences in the frequencies. The question is whether the quite large differences in Roald's case can be explained in this way or not.

When this is written in the formal language of statistical tests, it becomes:

H_0: $p = \frac{1}{6}$ for each outcome.

H_1: $p \neq \frac{1}{6}$ for each outcome.

The expected frequencies are denoted by f_e and the observed frequencies by f_o. To measure how far the observed data are from the expected, you clearly need to consider the difference between the observed frequencies, f_o, and the expected, f_e. The measure which is used as a test statistic for this is denoted by X^2 and given by:

$$X^2 = \sum_{\substack{\text{All} \\ \text{classes}}} \frac{(f_o - f_e)^2}{f_e}$$

In this case the calculation of X^2 is as follows:

Outcome	6	5	4	3	2	1
Observed frequency, f_o	12	16	15	23	24	30
Expected frequency, f_e	20	20	20	20	20	20
Difference, $f_o - f_e$	−8	−4	−5	3	4	10
$(f_o - f_e)^2$	64	16	25	9	16	100
$(f_o - f_e)^2/f_e$	3.2	0.8	1.25	0.45	0.8	5

$$X^2 = 3.2 + 0.8 + 1.25 + 0.45 + 0.8 + 5$$
$$= 11.5$$

The test statistic X^2 has the χ^2 (chi squared) distribution. Critical values for this distribution are given in tables but, before you can use them, you have to think about two more points.

WHAT IS TO BE THE SIGNIFICANCE LEVEL OF THE TEST?

This should really have been set before any data were collected. Because many people will be sceptical about Roald's claim, it would seem advisable to make the test rather strict, and so the 1% significance level is chosen.

HOW MANY DEGREES OF FREEDOM ARE INVOLVED?

You will recall from reading about the χ^2 distribution at the end of Chapter 2 that the shape of the curve depends on the number of free variables involved, the degrees of freedom, v.

In this case there are six classes (corresponding to scores on the die of 1, 2, 3, 4, 5 and 6) but since the total number of throws is fixed (120) the frequency in the last class can be worked out if you know those of the first five classes.

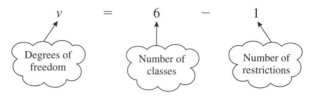

$$v \quad = \quad 6 \quad - \quad 1$$

Degrees of freedom   Number of classes   Number of restrictions

Looking up in the tables for the 1% significance level and $v = 5$ gives a critical value of 15.09; see figure 5.1.

Since $11.5 < 15.09$, H_0 is accepted.

There is no reason at this significance level to believe that any number on the die was any more likely to come up than any other. Roald's powers are not proved.

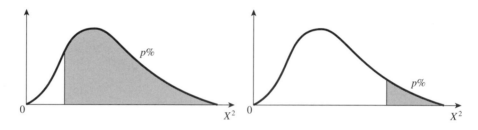

$p\%$	99	97.5	95	90		10	5.0	2.5	1.0	0.5
$v = 1$	.0001	.0010	.0039	.0158		2.706	3.841	5.024	6.635	7.879
2	.0201	.0506	0.103	0.211		4.605	5.991	7.378	9.210	10.60
3	0.115	0.216	0.352	0.584		6.251	7.815	9.348	11.34	12.84
4	0.297	0.484	0.711	1.064		7.779	9.488	11.14	13.28	14.86
5	0.554	0.831	1.145	1.610		9.236	11.07	12.83	15.09	16.75
6	0.872	1.237	1.635	2.204		10.64	12.59	14.45	16.81	18.55
7	1.239	1.690	2.167	2.833		12.02	14.07	16.01	18.48	20.28
8	1.646	2.180	2.733	3.490		13.36	15.51	17.53	20.09	21.95
9	2.088	2.700	3.325	4.168		14.68	16.92	19.02	21.67	23.59

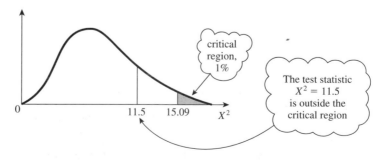

critical region, 1%

The test statistic $X^2 = 11.5$ is outside the critical region

Figure 5.1

Note

This is a one-tail test with only the right-hand tail under consideration. The interpretation of the left-hand tail (where the agreement seems to be too good) is discussed later in the chapter.

Properties of the test statistic X^2

$$X^2 = \sum_{\substack{\text{All} \\ \text{classes}}} \frac{(f_o - f_e)^2}{f_e}$$

- It is clear that as the difference between the expected values and the observed values increases then so will the value of this test statistic. Squaring the top gives due weight to any particularly large differences. It also means that all values are positive.
- Dividing $(f_e - f_o)^2$ by f_e has the effect of standardising that element, allowing for the fact that, the larger the expected frequency within a class, the larger will be the difference between the observed and the expected.

The χ^2 distribution with v degrees of freedom is that of

$$Z_1^2 + Z_2^2 + \ldots + Z_v^2$$

where $Z_1, Z_2, \ldots, Z_v$ are v independent standardised normal variables.

The detailed mathematics of how the expression

$$\sum_{\substack{\text{All} \\ \text{classes}}} \frac{(f_o - f_e)^2}{f_e}$$

fulfils this requirement is, however, beyond the scope of this book.

Notation

- An alternative notation which is often used is to call the expected frequency in the ith class E_i the observed frequency in the ith class O_i.

  In this notation, $X^2 = \sum_i \frac{(E_i - O_i)^2}{E_i}$
- The usual convention in statistics is to use a Greek letter for a parent population parameter and the corresponding Roman letter for the equivalent sample statistic.

Parent population parameters (Greek letters)		Values of sample statistics (Roman letters)
μ		m
σ		s
ρ	etc.	r

Unfortunately, when it comes to χ^2, there is no Roman equivalent to the Greek letter χ since it translates into CH. Since X looks rather like χ a sample statistic from a χ^2 population is denoted by X^2. (In the same way Christmas is abbreviated to χmas but written Xmas.)

Using the χ^2 test

The χ^2 test is commonly used to see if a proposed model fits observed data.

 Why is it the model that should fit the data and not the other way round?

Minimum expected frequencies

The expected frequency of any class must be at least five. If a class has an expected frequency of less than five, then it must be grouped together with one or more other classes until their combined expected frequency is at least five. In that case the number of classes is reduced accordingly when working out the degrees of freedom.

Degrees of freedom

The degrees of freedom for the test are given by:

$$\text{degrees of freedom} = \text{number of classes} - \text{number of restrictions}$$

In the case of Roald's die, there was one restriction, namely that the total of the frequencies had to be 120, the total number of throws. In some situations there are more restrictions than that, as you will see in the following examples.

When a particular distribution is fitted to the data, it may be necessary to estimate one or more parameters of the distribution. This, together with the restriction on the total, will reduce the number of degrees of freedom:

$$v = \text{number of classes} - \text{number of estimated parameters} - 1$$

EXAMPLE 5.1

Poisson distribution

The number of telephone calls made to a counselling service is thought to be modelled by the Poisson distribution. Data are collected on the number of calls received during one-hour periods as shown in the table. Use these data to test at the 5% significance level whether a Poisson model is appropriate.

No. of calls per hour	0	1	2	3	4	5	6	Total
Frequency	6	13	26	14	7	4	0	70

SOLUTION

H_0: The number of calls can be modelled by the Poisson distribution.

H_1: The number of calls cannot be modelled by the Poisson distribution.

Nothing is known about the form of the Poisson distribution, so the data must be used to estimate the Poisson parameter λ.

From the data, the mean number of calls per hour is

$$\frac{0 \times 6 + 1 \times 13 + 2 \times 26 + 3 \times 14 + 4 \times 7 + 5 \times 4}{70} = \frac{155}{70} = 2.214.$$

The Poisson distribution with parameter $\lambda = 2.214$ is as follows:

x	$P(X = x)$	$70 \times P(X = x)$	Expected frequency
0	$e^{-2.214}$	70×0.1093	7.6
1	$P(X = 0) \times \frac{2.214}{1}$	70×0.2419	16.9
2	$P(X = 1) \times \frac{2.214}{2}$	70×0.2678	18.7
3	$P(X = 2) \times \frac{2.214}{3}$	70×0.1976	13.8
4	$P(X = 3) \times \frac{2.214}{4}$	70×0.1094	7.7
5	$P(X = 4) \times \frac{2.214}{5}$	70×0.0484	3.4
$\geqslant 6$	$1 - P(X < 6)$	70×0.0256	1.8

The expected frequencies for the last two classes are both less than 5 but if they are put together to give an expected value of 5.2, the problem is overcome.

Note

The expected frequencies are not rounded to the nearest whole number. To do so would invalidate the test. Expected frequencies do not need to be integers.

The expected frequency for the last class was worked out as $1 - P(X < 6)$ and not as $P(X = 6)$, which would have cut off the right-hand tail of the distribution. The classes need to cover all *possible* outcomes, not just those that occurred in your survey.

The table for calculating the test statistic is shown below.

No. of calls, X	0	1	2	3	4	5+	Total
Observed frequency, f_o	6	13	26	14	7	4	70
Expected frequency, f_e	7.6	16.9	18.7	13.8	7.7	5.2	70
$(f_o - f_e)$	−1.6	−3.9	7.3	0.2	−0.7	−1.2	0.1
$(f_o - f_e)^2/f_e$		0.336	0.900	2.849	0.002	0.064	0.277	4.428

$X^2 = 0.336 + 0.900 + 2.849 + 0.002 + 0.064 + 0.277 = 4.428$

The degrees of freedom are

v = number of classes − number of estimated parameters − 1

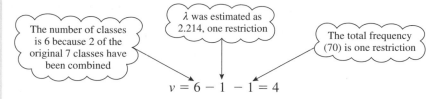

The number of classes is 6 because 2 of the original 7 classes have been combined

λ was estimated as 2.214, one restriction

The total frequency (70) is one restriction

$$v = 6 - 1 - 1 = 4$$

From the tables, the critical value for a significance level of 5% and 4 degrees of freedom is 9.488.

The calculated test statistic, $X^2 = 4.428$.

Since $4.428 < 9.488$, H_0 is accepted.

The data are consistent with a Poisson distribution for the number of calls.

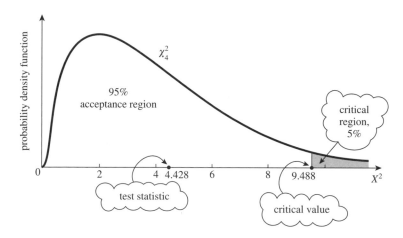

Figure 5.2

Note

The mode of χ^2_4 occurs at 2. The mode of χ^2_{10} occurs at 8. In general, the mode of χ^2_k occurs at $k - 2$.

EXAMPLE 5.2

Binomial distribution

An egg packaging firm has introduced a new box for its eggs. Each box holds six eggs. Unfortunately it finds that the new box tends to mark the eggs. Data on the number of eggs marked in 100 boxes are collected.

No. of marked eggs	0	1	2	3	4	5	6	**Total**
No. of boxes, f_o	3	3	27	29	10	7	21	100

It is thought that the distribution may be modelled by the binomial distribution. Carry out a test on the data at the 0.5% significance level to determine whether the data can be modelled by the binomial distribution.

SOLUTION

H_0: The number of marked eggs can be modelled by the binomial distribution.

H_1: The number of marked eggs cannot be modelled by the binomial distribution.

The binomial distribution has two parameters, n and p. The parameter n is clearly 6, but p is not known and so must be estimated from the data.

From the data, the mean number of marked eggs per box is

$$\frac{0 \times 3 + 1 \times 3 + 2 \times 27 + 3 \times 29 + 4 \times 10 + 5 \times 7 + 6 \times 21}{100} = 3.45.$$

Since the population mean is np we may estimate p by putting

$$6p = 3.45$$

$$\text{estimated } p = 0.575 \quad \text{and estimated} \quad q = 1 - p = 0.425.$$

These parameters are now used to calculate the expected frequencies of $0, 1, 2, \ldots, 6$ marked eggs per box in 100 boxes.

x	$P(X = x)$		Expected frequency, f_e $100 \times P(X = x)$
0	0.425^6	0.0059	0.59
1	$6 \times 0.575^1 \times 0.425^5$	0.0478	4.78
2	$15 \times 0.575^2 \times 0.425^4$	0.1618	16.18
3	$20 \times 0.575^3 \times 0.425^3$	0.2919	29.19
4	$15 \times 0.575^4 \times 0.425^2$	0.2962	29.62
5	$6 \times 0.575^5 \times 0.425^1$	0.1603	16.03
6	0.575^6	0.0361	3.61

In this case there are three classes with an expected frequency of less than 5. The class for $x = 0$ is combined with the class for $x = 1$, bringing the expected frequency just over 5, and the class for $x = 6$ is combined with that for $x = 5$.

No. of marked eggs, x	0, 1	2	3	4	5, 6	Total
Observed frequency, f_o	6	27	29	10	28	100
Expected frequency, f_e	5.37	16.18	29.19	29.62	19.64	100
$(f_o - f_e)$	0.68	10.82	−0.19	−19.62	8.36	
$(f_o - f_e)^2/f_e$	0.07	7.24	0.00	13.00	3.56	

The test statistic, $X^2 = 0.07 + 7.24 + 0.00 + 13.00 + 3.56 = 23.87$

The degrees of freedom,

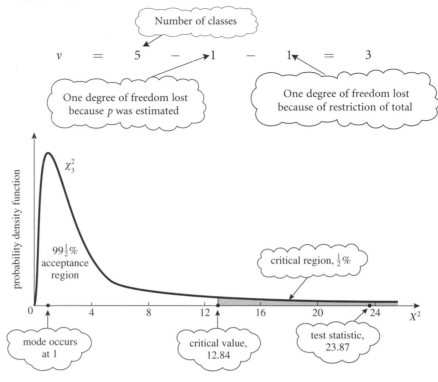

$$v \quad = \quad 5 \quad - \quad 1 \quad - \quad 1 \quad = \quad 3$$

Number of classes

One degree of freedom lost because p was estimated

One degree of freedom lost because of restriction of total

χ_3^2

probability density function

$99\frac{1}{2}\%$ acceptance region

critical region, $\frac{1}{2}\%$

mode occurs at 1

critical value, 12.84

test statistic, 23.87

X^2

Figure 5.3

From the tables, for the 0.5% significance level and $v = 3$, the critical value of χ^2 is 12.84.

Since $23.87 > 12.84$, H_0 is rejected.

The data indicate that the binomial distribution is not an appropriate model for the number of marked eggs. If you look at the distribution of the data you can easily see why, since it is bimodal (figure 5.4).

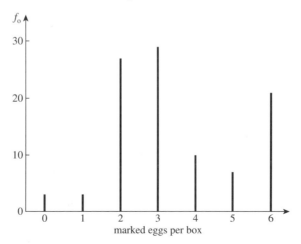

f_o

marked eggs per box

Figure 5.4

EXAMPLE 5.3

Given proportions

It is generally believed that a particular genetic defect is carried by 10% of people. A new and simple test becomes available to determine whether somebody is a carrier of this defect, using a blood specimen. As part of a research project, 100 hospitals are asked to carry out this test anonymously on the next 30 blood samples they take. The results are as follows:

Number of positive tests	0	1	2	3	4	5	6	7+
Frequency, f_o	11	29	26	20	9	3	1	1

Do these figures support the model that 10% of people carry this defect, independently of any other condition, at the 5% significance level?

SOLUTION

H_0: The model that 10% of people carry this defect is appropriate.

H_1: The model that 10% of people carry this defect is not appropriate.

The expected frequencies may be estimated using the binomial distribution $B(30, 0.1)$.

Number of positive tests	0	1	2	3	4	5	6	7+
Expected frequency, f_e	4.24	14.13	22.77	23.61	17.71	10.23	4.74	2.58

The calculation then proceeds as follows:

No. of positive tests	0, 1	2	3	4	5	6+	Total
Observed frequency, f_o	40	26	20	9	3	2	100
Expected frequency, f_e	18.37	22.77	23.61	17.71	10.23	7.32	100.01
$(f_o - f_e)$	21.63	3.23	−3.61	−8.71	−7.23	−5.32	
$(f_o - f_e)^2/f_e$	25.47	0.46	0.55	4.28	5.11	3.87	

The test statistic $X^2 = 25.47 + 0.46 + 0.55 + 4.28 + 5.11 + 3.87$

$$= 39.74$$

The degrees of freedom,

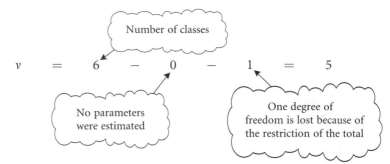

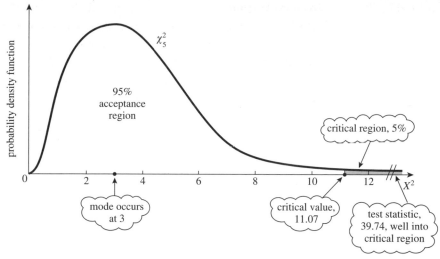

Figure 5.5

From the tables, for 5% significance level and $v = 5$, the critical value of χ^2 is 12.84.

$$\text{Since } 39.74 > 11.07, \text{ H}_0 \text{ is rejected.}$$

No, the data indicate that the binomial distribution with $p = 0.1$ is not an appropriate model.

Notes

1 Although this example is like the previous one in that both used the binomial distribution as a model, the procedure is different. In this case the given model included the information $p = 0.1$ and so you did not have to estimate the parameter p. Consequently a degree of freedom was not lost from doing so.

The model B(30, 0.1) specified the probability for each of the classes. This is an example of a 'given' proportions model.

2 The value of X^2 was very large in comparison with the critical value. What went wrong with the model?

You will find that if you use the data to estimate p, it does not work out to be 0.1 but a little under 0.07. Fewer people are carriers of the defect than was believed to be the case. If you work through the example again with the model $p = 0.07$, you will find that the fit is good enough for you to start looking at the left-hand tail of the distribution.

The left-hand tail

The χ^2 test is conducted as a one-tail test, looking to see if the test statistic gives a value to the right of the critical value, as in the previous examples.

However, examination of the left-hand tail also gives information. In any modelling situation you would expect there to be some variability. Even when using the binomial to model a clear binomial situation, like the number of heads obtained in throwing a coin a large number of times, you would be very surprised if the observed and expected frequencies were identical. The left-hand tail may lead you to wonder whether the fit is too good to be credible. The following is a very famous example of the Poisson distribution.

Death from horse kicks

For a period of 20 years in the nineteenth century data were collected of the annual number of deaths caused by horse kicks per army corps in the Prussian army.

No. of deaths	0	1	2	3	4
No. of corps, f_o	109	65	22	3	1

These data give a mean of 0.61. The variance of 0.6079 is almost the same, suggesting that the Poisson model may be appropriate.

The distribution Poisson (0.61) gives these figures:

No. of deaths	0	1	2	3	4 or more
No. of corps, f_e	108.7	66.3	20.2	4.1	0.7

This looks so close that there seems little point in using a test to see if the data will fit the distribution. However, proceeding to the test, and remembering to combine classes to give expected values of at least 5, the null and alternative hypotheses are as follows.

H_0: Deaths from horse kick can be modelled by a Poisson distribution.

H_1: Deaths from horse kick cannot be modelled by a Poisson distribution.

No. of deaths	0	1	2 or more
Observed frequency, f_o	109	65	26
Expected frequency, f_e	108.7	66.3	25
$(f_o - f_e)$	0.3	-1.3	1
$(f_o - f_e)^2/f_e$	0.001	0.025	0.040

The test statistic $X^2 = 0.001 + 0.025 + 0.040$

$$= 0.066$$

The degrees of freedom are given by

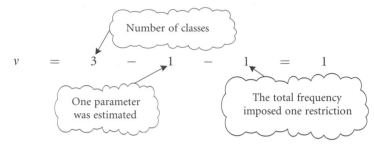

The critical value of χ^2 for 1 degree of freedom at the 5% significance level is 3.841.

$$0.066 < 3.841$$

It is clear that the null hypothesis, that the Poisson distribution is an appropriate model, should be accepted.

? Looking at the expected values of the Poisson distribution in comparison with the observed values suggests that the fit is very good indeed. Is it perhaps suspiciously good? Might the data have been fixed?

The tables relating to the left-hand tail of the χ^2 distribution give critical values for this situation. For example the value for 95% significance level for $v = 1$ is 0.0039. This means if the null hypothesis is true you would expect a value for X^2 less than 0.0039 from no more than 5% of samples.

In this case, the test statistic is $X^2 = 0.066$. This is greater than the 95% critical value, and so you can conclude that a fit as good as this will occur with more than 1 sample in 20. That may well help allay your suspicions.

If your test statistic does lie within the left-hand critical region, you should check the data to ensure that the figures are genuine and that all the procedures have been carried out properly. There are three situations you should particularly watch out for.

- The model was constructed to fit a set of data. It is then being tested by seeing how well it fits the same data. Once the model is determined, new data should be used to test it.

- Some of the data have been omitted in order to produce a better fit.

- The data are not genuine.

Although looking at the left-hand tail of the χ^2 distribution may make you suspicious of the quality of the data, it does not provide a formal hypothesis test that the data are not genuine. Thus the term *critical region* is not really appropriate to this tail; *warning region* would be better.

5

Continuous distributions

The χ^2 test can be used just as easily with continuous distributions as it can with discrete distributions. However, calculating the expected values may involve a little more work. This is illustrated in the following example using the normal distribution.

EXAMPLE 5.4

The manager of a workshop has introduced a scheme to reduce the waste offcut of hardwood timber. The manager suspects that the distribution of lengths of waste timber can be modelled by the normal distribution. 100 lengths of waste timber are measured as follows:

Length, l (mm)	$l < 20$	$20 \leqslant l < 40$	$40 \leqslant l < 60$	$60 \leqslant l < 80$	$80 \leqslant l < 100$	$l \geqslant 100$
Frequency	5	16	33	23	15	8

Carry out a test at the 5% significance level of whether this model is suitable.

SOLUTION

H_0: The lengths of waste timber can be modelled by the normal distribution.

H_1: The lengths of waste timber cannot be modelled by the normal distribution.

In order to fit a normal distribution to these data it is necessary to know two parameters, the mean and the standard deviation.

Assuming that the shortest length of timber is 0 mm (obviously!) and the longest is 120 mm, the mean and standard deviation are estimated from the data.

Using $$\bar{x} = \sum_i \frac{x_i}{n} \quad \text{and} \quad s^2 = \sum_i \frac{(x_i - x)^2}{n - 1}$$

gives estimate of $\mu = \bar{x} = 60.2$ and estimate of $\sigma = s = 25.819$.

Note

Two parameters have been estimated here and so the degrees of freedom are reduced by two.

The upper class boundaries are 20, 40, 60, 80, 100, (the last class is open-ended). These must be standardised in the usual way, and then the relevant figures looked up in normal distribution tables.

Class	Upper bound	Standardised upper bound	$P(Z < z)$	Probability for class	Expected frequency
$l < 20$	20	−1.557	0.0597	0.0597	5.97
$20 \leqslant l < 40$	40	−0.0782	0.2171	0.1574	15.74
$40 \leqslant l < 60$	60	−0.008	0.4968	0.2797	27.97
$60 \leqslant l < 80$	80	0.767	0.7785	0.2817	28.17
$80 \leqslant l < 100$	100	1.541	0.9383	0.1598	15.98
$l \geqslant 100$	—	—	1.0000	0.0617	6.17

The figures for the expected frequency show that there is no need to put any classes together as they are all greater than 5. The usual table can now be set up to calculate the test statistic, X^2.

	< 20	20–40	40–60	60–80	80–100	$\geqslant 100$	Total
Observed frequency, f_o	5	16	33	23	15	8	100
Expected frequency, f_e	5.97	15.74	27.97	28.17	15.98	6.17	100
$(f_o - f_e)$		−0.97	0.26	5.03	−5.17	−0.98	1.83
$(f_o - f_e)^2/f_e$		0.158	0.004	0.905	0.949	0.060	0.543

$$X^2 = 0.158 + 0.004 + 0.905 + 0.949 + 0.060 + 0.543 = 2.618$$

The degrees of freedom,

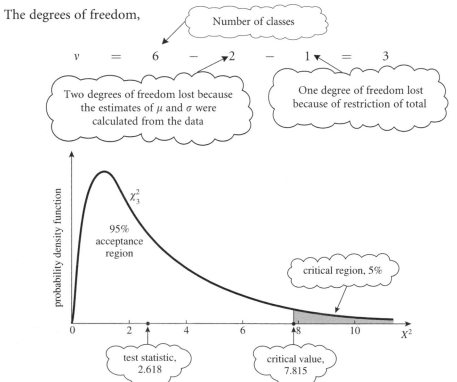

Figure 5.6

For $v = 3$ at the 5% significance level, the critical value of χ^2 is 7.815.

Since $2.618 < 7.815$, H_0 is accepted.

The data indicate that the normal distribution is an appropriate model.

Note

The χ^2 test is a distribution-free test; that means that there are no modelling assumptions associated with the test itself. This has the advantage that the test can be widely used, but on the other hand it is not a very sensitive test. (The test for rank correlation is another example of a distribution-free test: one can test for rank correlation without knowing anything about the underlying bivariate distribution, but one frequently finds there is no significant evidence of correlation.)

EXERCISE 5A

All the questions in this exercise require you to carry out a hypothesis test. The answer to each question must contain a clear statement of your null and alternative hypotheses and the conclusion drawn from the test.

1 A typist makes mistakes from time to time in a 200-page book. The number of pages with different numbers of mistakes are as follows:

Mistakes	0	1	2	3	4	5
Pages	18	62	84	30	5	1

(i) Test at the 5% significance level whether the Poisson distribution is an appropriate model for these data.

(ii) What factors would make it other than a Poisson distribution?

2 A biologist crosses two pure varieties of plant, one with pink flowers, the other white. The pink is dominant so that the flowers of the second generation should be in the ratio

pink : white $= 3:1$.

He plants the seeds in batches of 5 in 32 trays and counts the numbers of plants with pink and with white flowers in each tray.

White flowers	0	1	2	3	4	5
Frequency	9	12	7	2	1	1

(i) What distribution would you expect for the number of plants with white flowers?

(ii) Use these figures to test at the 2.5% significance level whether the distribution is that which you expected.

3 A man accuses a casino of having two loaded dice. He throws them 360 times with the following outcomes for their sum at each throw:

Total	2	3	4	5	6	7	8	9	10	11	12
Freq.	5	15	30	35	45	61	53	45	31	24	16

Does he have grounds for his accusation at the 5% significance level?

4 An examination board is testing a multiple-choice question. They get 100 students to try the question and their answers are as follows:

Choice	A	B	C	D	E
Frequency	32	18	10	28	12

Are there grounds, at the 10% significance level, for the view that the question was so hard that the students guessed the answers at random?

5 Between 1989 and 1998 a mathematics department recorded its A level grades, and again in 1999:

	A	B	C	D	E	N	U
1989 to 1998	50	71	119	65	35	18	12
1999	3	5	11	7	5	3	0

The teachers believed that the 1989 to 1998 figures were typical and used them to formulate a prediction model: that in subsequent years the same proportions of candidates would get the various grades.

Carry out a test at the 5% significance level to determine whether this model was appropriate for the 1999 candidates.

6 A student on a geography field trip has collected data on the size of rocks found on a scree slope. The student counts the number of large rocks (that is, heavier than a stated weight) found in a 2 m square at the top, middle and bottom of the slope.

	Top	Middle	Bottom
Number of large rocks	5	10	18

(i) Test at the 5% significance level whether these data are consistent with the hypothesis that the size of rocks is distributed evenly on the scree slope.

(ii) What does your test tell the student about the theory that large rocks will migrate to the bottom of the slope?

7 A regular gambler at a casino thinks that the roulette wheel is biased. The wheel has 37 equal sectors, each of which is given a number between 0 and 36. The number 0 is coloured green and the other numbers are equally divided between red and black. You can bet on which colour sector the ball will settle in when the wheel stops turning. At the suggestion of the management, the gambler records the number of times each of these three colours occurs during an evening's gambling. The results are:

	Green	Red	Black	Total
Frequency	28	325	387	740

Carry out a test at the 1% significance level to see if the gambler is justified in his allegation that the wheel is biased.

8 A university student working in a small seaside hotel in the summer holidays looks at the records for the previous holiday season of 30 weeks. She records the number of days in each week on which the hotel had to turn away visitors because it was full. The data she collects are:

Number of days visitors turned away	0	1	2	3	4	5+	Total
Number of weeks	11	13	4	1	1	0	30

(i) Calculate the mean and variance of the data.

(ii) The student thinks that these data can be modelled by the binomial distribution. Carry out a test at the 5% significance level to see if the binomial distribution is a suitable model.

(iii) What other distribution might be used to model these data? Give your reasons.

9 Morag is writing a book. Every so often she uses the spell check facility in her word processing software, and for interest records the number of mistakes she has made on each page. In the first 20 pages the results were as follows:

No. of mistakes/page	0	1	2	3	4+	Total
Frequency	9	6	4	1	0	20

(i) Explain why it is not possible to use the χ^2 test on these data to decide whether the occurrence of spelling mistakes may be modelled by the Poisson distribution.

In the next 30 pages Morag's figures are as follows:

No. of mistakes/page	0	1	2	3	4+	Total
Frequency	14	7	7	0	2	30

(ii) Use the combined figures, covering the first 50 pages, to test whether the occurrence of Morag's spelling mistakes may be modelled by the Poisson distribution. Use the 5% significance level.

(iii) If the distribution really is Poisson, what does this tell you about the incidence of spelling mistakes? Do you think this is realistic?

10 The head of a large computing department keeps a record of the number of computers that are not working on any particular day.

Computers not working	0	1	2	3	4	5	Total
Frequency	44	27	16	8	5	0	100

In order to decide how many spare computers to keep, he wishes to model this situation using the Poisson distribution. Construct and carry out a test to see if he is justified in using this distribution.

11 The research department of a soft drinks manufacturer has come up with a new recipe for one of its brands. The replacement is cheaper to produce and the company proposes introducing it without telling the public. Consequently the company needs to know whether people will notice the difference. A blind tasting is arranged and 20 people are given both drinks and asked to decide which is the old one and which the new. Fifteen choose correctly.

(i) Carry out a χ^2 test to see if the new drink is distinguishable. State clearly your null and alternative hypotheses and the significance level you are using.

An alternative test procedure is to use a test based on the binomial distribution rather than the χ^2 distribution.

(ii) Carry out this procedure and compare the result.

(iii) Discuss whether the sampling procedure will provide the company with the information it requires.

12 In a survey of five towns the population of the town and the number of petrol-filling stations were recorded as follows:

Town	Population (to nearest 10 000)	Number of filling stations
A	4	22
B	3	16
C	7	35
D	6	27
E	12	60
Totals	32	160

(i) An assistant researcher, who wanted to find out whether the petrol stations were evenly distributed between the towns, performed a χ^2 test on the number of filling stations, with a null hypothesis that there was no difference in the number of filling stations in each town. She found that her X^2 value was 36.69. Without repeating her calculation, state with reasons what her conclusion was.

(ii) The senior researcher decided to use the hypothesis that the number of filling stations was directly proportional to the size of the town's population. Show that, on this basis, the expected number of garages for town A would be 20, and test the hypothesis at the 10% significance level stating your conclusions clearly.

[MEI]

13 A local council has records of the number of children and the number of households in its area. It is therefore known that the average number of children per household is 1.40. It is suggested that the number of children per household can be modelled by the Poisson distribution with parameter 1.40. In order to test this, a random sample of 1000 households is taken, giving the following data.

Number of children	0	1	2	3	4	5+
Number of households	273	361	263	78	21	4

(i) Find the corresponding expected frequencies obtained from the Poisson distribution with parameter 1.40.

(ii) Carry out a χ^2 test, at the 5% level of significance, to determine whether or not the proposed model should be accepted. State clearly the null and alternative hypotheses being tested and the conclusion which is reached.

[MEI]

14 (i) A random sample of supermarkets were sent a questionnaire on which they were asked to report the number of cases of shoplifting they had dealt with in each month of the previous year. The totals for each month were as follows.

J	F	M	A	M	J	J	A	S	O	N	D
16	12	10	17	6	18	16	17	10	22	14	16

Carry out a χ^2 test at an appropriate level of significance to determine whether or not shoplifting is more likely to occur in some months than others. (You may take all months to be of the same length.) Make clear your null and alternative hypotheses, the level of significance you are using, and your conclusion.

You may, if you wish, use the fact that, when all the values of f_e are equal, the usual χ^2 test statistic may be written as

$$\frac{1}{f_e}\sum f_o^2 - \sum f_o$$

(ii) Prove the result given at the end of part (i).

[MEI]

15 In a survey 200 motorists, chosen at random, were asked how many attempts they took at the driving test until they passed. The results are summarised by the following frequency distribution:

Number of attempts	1	2	3	4	5	6	7	8+
Number of motorists	71	47	35	20	15	9	3	0

(i) Calculate the sample mean $\bar{x}$.

A statistician suggests that the number of attempts required to pass the driving test can be modelled by the random variable X defined as follows:

$$P(X = r) = p(1 - p)^{r-1} \quad \text{for} \quad r = 1, 2, 3, \ldots$$

where p represents the probability of a motorist passing any particular test.

You are given that, for this model, $E(X) = \dfrac{1}{p}$, so p can be estimated by $\dfrac{1}{\bar{x}}$.

(ii) Calculate the expected frequencies according to the statistician's model.
(iii) Carry out an appropriate test, at the 5% level, to determine whether the model is a good one or not. State your hypotheses and conclusions carefully, and justify the number of degrees of freedom used.

[MEI]

16 Properties where people live are placed in one of eight bands for council tax assessment. Band A represents the lowest-value properties (up to £40 000) and band H represents the highest-value properties (over £320 000).

It is desired to investigate whether the distribution of properties within the eight bands in the city of Trumpton matches the national distribution.

A random sample of 500 properties in Trumpton is taken. The percentage of properties in each band nationally and the number of properties in each band in the sample are as follows:

Band	Percentage of properties nationally	Number of properties in sample
A	14.2%	53
B	22.6%	105
C	20.8%	111
D	14.3%	80
E	11.4%	63
F	9.8%	49
G	5.9%	36
H	1.0%	3
Total	100.0%	500

(i) Find the expected frequency within each band in Trumpton based on the national proportions.

(ii) Carry out an appropriate hypothesis test, at the 10% significance level. State your hypotheses and conclusions carefully.

(iii) If a random sample of 1000 properties had been taken and there had been precisely *twice* as many properties in each band, what effect would this have on your conclusions in part (ii)?

[MEI]

17 A garden on a Mediterranean island is populated by lizards which come to feed if scraps of food are left for them. The number of lizards feeding in the garden is observed at noon on each of 70 days. The numbers of days f on which n lizards are observed feeding in the garden are recorded in the table below.

n	0	1	2	3	4 or more
f	7	16	20	15	12

(i) Calculate, correct to 2 decimal places, expected frequencies for each of the classes in the table, assuming that the number of lizards observed has a Poisson distribution with mean 2.

(ii) Carry out a χ^2 goodness-of-fit test to test, at the 10% significance level, the null hypothesis that the number of lizards observed has a Poisson distribution with mean 2, against the alternative hypothesis that this is not the case.

[Cambridge]

18 The height, in centimetres, gained by a conifer in its first year after planting is denoted by the random variable X. The value of X is measured for a random sample of 86 conifers and the results obtained are summarised in the table.

X	< 35	35–45	45–55	55–65	> 65
Observed frequency	10	18	28	18	12

(i) Assuming that the random variable X is modelled by a $N(50, 15^2)$ distribution, calculate the expected frequencies for each of the five classes.

(ii) Carry out a χ^2 goodness-of-fit analysis to test, at the 5% significance level, the hypothesis that X can be modelled as in part (i).

[Cambridge]

INVESTIGATIONS

1 Select 100 digits from random number tables (or from your calculator) and carry out a test at the 1% significance level to determine whether the selection is in fact truly random. What aspects of randomness does your test fail to detect?

2 Throw a six-sided die 48 times and record your results in a frequency table. Conduct a hypothesis test at the 10% significance level to determine whether the frequency distribution you obtain is that of a fair die.

KEY POINTS

The steps in using the χ^2 test are as follows.

1 Select your model.
This means deciding which distribution is to be used to model the situation. It may be a standard distribution like the binomial, or it may just be some known probabilities.

2 Set up null and alternative hypotheses and choose the significance level.
Since the χ^2 test is one-tailed, the null hypothesis will be that the data are drawn from the model population, the alternative hypothesis that they are not. Choose a significance level appropriate to the situation.

3 Collect data.
The frequencies in the different classes are denoted by f_o.

4 Calculate the expected frequencies.
These are generated by the distribution specified in the null hypothesis, and denoted by f_e. The values of f_e should not be rounded.

5 **Check the sizes of the classes.**
Ensure the data are grouped so that the expected frequency in each of the classes is five or more.

6 **Calculate the test statistic.**
This is denoted by X^2 and given by

$$X^2 = \sum_i \frac{(f_o - f_e)^2}{f_e}$$

7 **Work out the degrees of freedom.**
This is denoted by v and given by

$$v = \text{number of classes} - \text{number of restrictions}$$

which may alternatively be written

$$v = \text{number of classes} - \text{number of estimated parameters} - 1$$

8 **Find the critical value.**
Read the critical value from the χ^2 tables for the number of degrees of freedom and the required significance level.

9 **Carry out the test.**
Compare the test statistic, X^2, with the critical value.

If $X^2 <$ critical value then accept H_0,

if $X^2 >$ critical value then reject H_0.

10 **Draw conclusions from the test.**
State what the test tells you about your model.

Answers

Chapter 1

❓ (Page 3)

It is reasonable to regard the height of a wave as random. No two waves are exactly the same and in a storm some are much bigger than others.

Exercise 1A (Page 9)

1 **(i)** $k = \frac{2}{35}$

(ii)

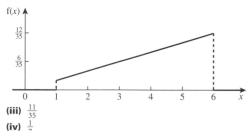

(iii) $\frac{11}{35}$

(iv) $\frac{1}{7}$

2 **(i)** $k = \frac{1}{12}$

(ii)

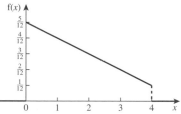

(iii) 0.207

3 **(i)** $a = \frac{4}{81}$

(ii)

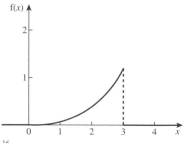

(iii) $\frac{16}{81}$

4 **(i)** $k = \frac{1}{4}$

(ii)

(iii) $\frac{27}{32}$

5 **(i)** $c = \frac{1}{8}$

(ii)

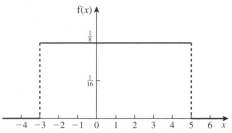

(iii) $\frac{1}{4}$

(iv) $\frac{3}{8}$

6 **(i)** $k = 0.048$

(ii)

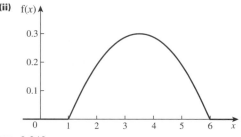

(iii) 0.248

7 **(i)** $a = \frac{5}{12}$

(ii)

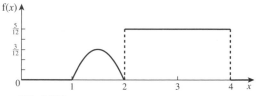

(iii) 0.292

(iv) $\frac{7}{12}$

8 **(i)** $k = \frac{2}{9}$

(ii) 0.067

9 **(i)** $k = \frac{1}{100}$

(ii)

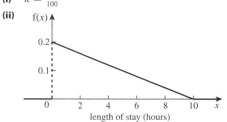

(iii) 19, 17, 28, 36

(iv) Yes

(v) Further information needed about the group 4–10 hours. It is possible that many of these stay all day are so are part of a different distribution.

10 (i)

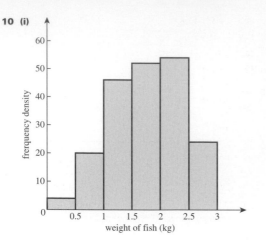

Negative skew

(ii)

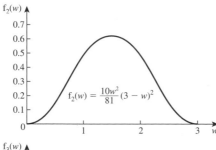

$f_1(w) = \frac{2w}{9}(3 - w)$

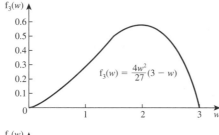

$f_2(w) = \frac{10w^2}{81}(3 - w)^2$

$f_3(w) = \frac{4w^2}{27}(3 - w)$

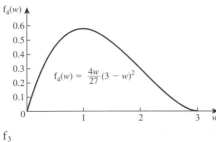

$f_4(w) = \frac{4w}{27}(3 - w)^2$

f_3

(iii) 1.62, 9.49, 20.14, 28.01, 27.55, 13.19

(iv) Model seems good.

11 (i) $a = 100$ **(ii)** 0.045   **(iii)** 0.36

12 0.803, 0.456

---

13 (i) 0, 0.1, 0.21, 0.12, 0.05, 0.02, 0

    (ii) 0.1, 0.31, 0.33, 0.17, 0.07, 0.02

    (iii) $k = \frac{1}{1728}$

    (iv) 0.132, 0.275, 0.280, 0.201, 0.095, 0.016

    (v) Model quite good. Both positively skewed.

? (Page 16)

(b) and (d)

? (Page 20)

68%. The normal distribution has a greater proportion of values near the mean, as can be seen from its shape.

Exercise 1B (Page 23)

1 (i) 2.67

    (ii) 0.89

    (iii) 2.828

2 (i) 2   **(ii)** 2   **(iii)** 1.76

3 (i) 0.6

    (ii) 0.04

    (iii) $\frac{2}{3}$

4 (i)

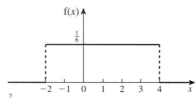

    (ii) $\frac{2}{3}$

    (iii) 1

    (iv) $\frac{1}{3}$

5 (i) 1.5

    (ii) 0.45

    (iii) 1.5

    (iv) 1.5

    (v)

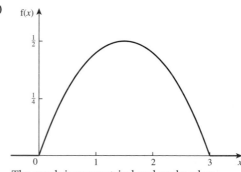

The graph is symmetrical and peaks when $x = 1.5$ thus $E(X) =$ mode of X = median value of $X = 1.5$.

6 (i) $f(x) = \frac{1}{7}$ for $-2 \leqslant x \leqslant 5$

    (ii) 1.5

    (iii) 4.08

    (iv) $\frac{5}{7}$

7 (ii) 1.083, 0.326

(iii) 0.5625

8 (i) $f(x) = \frac{1}{3}$ for $4 \leqslant x \leqslant 7$

(ii) 5.5

(iii) $\frac{3}{4}$

(iv) 0.233

9 (i) $f(x) = \frac{1}{10}$ for $10 \leqslant x \leqslant 20$

(ii) 15, 8.33

(iii) (a) 57.7%

(b) 100%

10 (i)

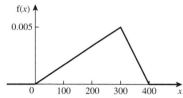

(iii) 233 hours

(iv) 7222.2

(v) 0.083

11 (i) $k = 1.2 \times 10^{-8}$

(ii)

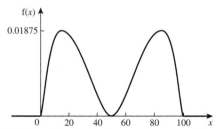

(iii) The distribution is the sum of two smaller distributions, one of moderate candidates and the other of able ones.

(iv) Yes if the step size is small compared to the standard deviation.

12 (ii) 8.88, 2.88; 0.724

(iii) $2m^3 - 18m^2 + 78m - 900 = 0$

13 (i) $a = k$

(ii) $\frac{1}{k}$

(iii) $\frac{1}{k^2}$

(iv) $\frac{\ln 2}{k}$

(v) Life length in hours of an electric light bulb; $k = \frac{1}{200}$

14 (i) 200

(ii) 0.082

(iii) 0.139

(iv) $k = 7.31$

15 (i) $a = 1.443$

(ii)

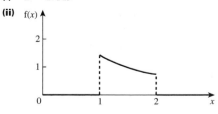

(iii) 1.443, 0.083

(iv) 41.5%

(v) 1.414

16 (i)

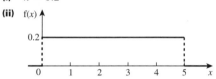

exam mark

The model suggests that these candidates were generally of high ability as a large proportion of them scored a high mark.

(iii) 12.5%

(iv) No; 91

Exercise 1C (Page 30)

1 0.8, 0.16, £8

2 (i) $k = 0.2$

(ii)

(iii) 2.5

(iv) 7

3 (i) 0.8

(ii) $0.\dot{6}$

(iii) $0.02\dot{6}$

4 (i) 1.5

(ii) 2.7

(iii) 0.45

(iv) 13.9

(v) 0.45; both are the variance of Y.

5 (i) 0.6

(ii) -3.4

(iii) 0.2

(iv) 0.64

6 (i) $3\frac{2}{3}$

(ii) $66\frac{1}{6}$

(iii) $14\frac{5}{6}$, $66\frac{1}{6}$

7 (i) $f(x) = \frac{1}{6}$ for $2 \leqslant x \leqslant 8$

(ii) $a = \frac{x^2\sqrt{3}}{4}$

(iii) 0.352

(iv) 12.12, 57.6

8 $k = \frac{3}{32}$, 2; $\frac{5}{32}$

9 (i) (a) 100, 11.55

(b) 1013.3, 231.2

(ii) (a) $f(u) = \frac{1}{400}u - \frac{1}{5}$, $80 \leqslant u \leqslant 100$

$f(u) = \frac{3}{10} - \frac{1}{400}u$, $100 \leqslant u \leqslant 120$

(b) 100, 66.7

(c) 1007, 26 729

10 (i) 100 days

(ii) 0.026

(iii) 300 days

(iv) £75 000

(v) £8333

11 (i) $f(t) = 0.1$ for $0 \leqslant t \leqslant 10$

       $= 0$ otherwise

      mean $= 5$, variance $= 8\frac{1}{3}$

(ii)

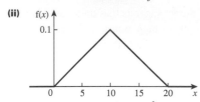

(iii) $E(X) = 10$, $\text{Var}(X) = 16\frac{2}{3}$

(iv) 0.18; because $T_1 \geqslant 7$ and $T_2 \geqslant 7$ is not the only way for $X \geqslant 14$. The latter inequality also includes other possibilities, such as waiting 9 minutes in the morning and 6 minutes in the evening.

12 (i) $E(X) = 3.2$

(ii)

p.d.f.

time (minutes)

The model implies that all of the doctor's appointments last between 2 and 10 minutes, the mean time being 5.2 minutes and the variance of the distribution being 2.56 minutes^2.

❓ (Page 36)

The model's predictions agree closely with the actual data. However, the value of the model lies in how well it will predict the pattern for future years. With a large entry the pattern should be reasonably consistent from year to year. This model is not very good from $1\frac{1}{2}$ to 2 hours, when more people finish than it predicts.

Exercise 1D (Page 47)

1 (i) 2.5

(ii) $F(x) = 0$ for $x < 0$

       $\dfrac{x}{5}$ for $0 \leqslant x \leqslant 5$

       1 for $x > 5$

(iii) 0.4

2 (i) $k = \frac{2}{39}$

(ii) f(u)

(iii) $F(u) = 0$ for $u < 5$

      $= \dfrac{u^2}{39} - \dfrac{25}{39}$ for $5 \leqslant u \leqslant 8$

      $= 1$ for $u > 8$

(iv) F(u)

3 (i) $c = \frac{1}{21}$

(ii) $F(x) = 0$ for $x < 1$

      $= \dfrac{x^3}{63} - \dfrac{1}{63}$ for $1 \leqslant x \leqslant 4$

      $= 1$ for $x > 4$

(iii) 3.19

(iv) 4

4 (i)

      $F(x) = 0$ for $x < 0$

      $= 1 - \dfrac{1}{(1+x)^3}$ for $x \geqslant 0$;

(ii) $x = 1$

5 (i) $\frac{1}{4}$

(ii) 0.134

(iii) $f(x) = 2 - 2x$ for $0 \leqslant x \leqslant 1$

f(x)

6 $E(X) = \frac{3}{4}$, $\text{Var}(X) = \frac{19}{80}$

   $F(x) = 0$ for $x < 0$

      $= \dfrac{3x}{4} - \dfrac{x^3}{16}$ for $0 \leqslant x \leqslant 2$

      $= 1$ for $x > 2$

7 (i) $\frac{3}{5}$, 0.683

8 (ii) $F(t) = 0$ for $t < 0$

      $= \dfrac{t^3}{432} - \dfrac{t^4}{6912}$ for $0 \leqslant t \leqslant 12$

      $= 1$ for $t > 12$

(iv) 0.132

9 (i) 2.93

(ii) $F(x) = 1 - \dfrac{(x - 10)^2}{100}$ for $0 \leqslant x \leqslant 10$

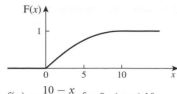

(iii) $f(x) = \dfrac{10 - x}{50}$ for $0 \leqslant x \leqslant 10$

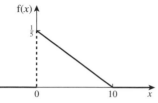

10 $F(x) = \dfrac{k}{2} - \dfrac{k\cos2x}{2}$; 0.146

11 (i) $\frac{1}{3}$

(iii) 4.39

(iv) 12.5

12 (i) (a) 0.3935

(b) 0.2231

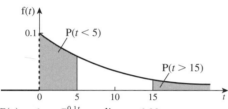

(ii) $F(t) = 1 - e^{-0.1t}$; median = 6.93

(iii) 0.183

13 (i) $1 - \dfrac{1}{m} = 0.4$, $m = 1.67$

(iii) 0.495

(iv) $f(x) = \dfrac{1.25}{x^2}$ for $1 \leqslant x \leqslant 5$

$= 0$ otherwise

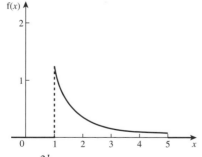

(v) $m = \dfrac{2b}{b + 1}$, which is always less than 2

14 $\mu = \dfrac{a}{2}$, $\sigma = \dfrac{a}{2\sqrt{3}}$

$f(y) = \dfrac{y}{a^2}$ for $0 \leqslant y \leqslant a$

$= \dfrac{2a - y}{a^2}$ for $a \leqslant y \leqslant 2a$

15 $F(x) = \int f(x)\,dx$ and so $F'(x) = f(x)$.

Uniform distribution with mean value $\dfrac{a}{2}$.

$F(x) = 1 - \dfrac{(a - x)^2}{a^2}$ for $0 \leqslant x \leqslant a$

Mean of sum of smaller parts $= a$

16 (i) (a) Validates p.d.f. form of Z.

(b) Demonstrates that $E(Z) = 0$.

(ii) (a) $E(Y) = 1$

(b) $Var(Y) = 2$

17 (i) $\{y : y \geqslant 0\}$

(ii) If $Y \leqslant y$ then either $0 \leqslant Z \leqslant \sqrt{y}$

or $-\sqrt{y} \leqslant Z \leqslant 0$

$\therefore P(Y \leqslant y)$

$= P(0 \leqslant Z \leqslant \sqrt{y}) + P(-\sqrt{y} \leqslant Z \leqslant 0)$

$= 2P(0 \leqslant Z \leqslant \sqrt{y})$

since Z is an even function.

Chapter 2

Exercise 2A (Page 58)

1 (i) 4, 0.875

(ii) 1.5, 0.167

(iii)

Main course	Dessert	Price
Fish and chips	Ice cream	£4
Fish and chips	Apple pie	£4.50
Fish and chips	Sponge pudding	£5
Bacon and eggs	Ice cream	£4.50
Bacon and eggs	Apple pie	£5
Bacon and eggs	Sponge pudding	£5.50
Pizza	Ice cream	£5
Pizza	Apple pie	£5.50
Pizza	Sponge pudding	£6
Steak and chips	Ice cream	£6.50
Steak and chips	Apple pie	£7
Steak and chips	Sponge pudding	£7.50

(iv) Mean of $T = 5.5$, variance $= 1.042$

2 (i) N(90, 25)

(ii) N(10, 25)

(iii) N(−10, 25)

3 0.196

4 (i) 0.0228

(ii) 56.45 minutes

(iii) 0.362

5 (i) 230 g, 9.8 g

(ii) 0.1587

(iii) 0.0787

6 (i) N(70, 25)

(ii) N(−10, 25)

7 5.92%

8 (i) 0.266

(ii) No, people do not choose their spouses at random: the height of a husband and wife may not be independent.

9 0.151

10 (i) 0

(ii) 0.0037

❓ (Page 63)

With folded paper it is not possible for pieces of paper that are thicker to be offset by others that are thinner, and vice versa.

Exercise 2B (Page 63)

1 $N(120, 24)$

Assume times are independent and no time is spent on changeovers between vehicles.

2 0.0745

3 0.1377

4 0.1946

5 (i) $N(34, 30)$

(ii) $N(-4, 30)$

(iii) $N(24, 29)$

6 (i) 0.316

(ii) 0.316

7 (i) $N(100, 26)$

(ii) $N(295, 353)$

(iii) $N(200, 122)$

(iv) $N(-65, 377)$

8 (i) 0.0827

(ii) 0.3103

(iii) 0.5

9 (i) 0.0827

(ii) 0.1446

(iii) 0.5

(iv) The situations in 8(i) and 9(i) are the same. 8(ii) considers $3X + 5Y$ whereas 9(ii) considers $X_1 + X_2 + X_3 + Y_1 + \ldots + Y_5$, so the probabilities are different.

8(iii) and 9(iii) have the same probabilities because $\dfrac{0}{\sigma} = 0$, so the variance does not affect the answer.

10 (i) $N(7400, 28\,900)$

(ii) $N(1200, 27\,700)$

(iii) $N(600a + 1000b, 400a^2 + 900b^2)$

11 (i) 311.6 kg

(ii) Assume that the composition of each crew is selected randomly so that the weights of each of the four individual oarsmen are independent

of each other. This assumption may not be reasonable since there may be some light-weight and some heavy-weight crews; also men's and women's crews. If this is so it will cast doubt on the answer to part (i).

12 (i) 0.4546

(ii) 93.491

13 (i) 0.234

(ii) 0.6915

14 0.9026

Assume weights of participants are independent since told teams were chosen at random.

15 (i) 0.0188

(ii) 0.1394

16 (i) 0.0367

(ii) 0.8144

(iii) 108, 1.4

17 (i) $N(80, 8)$

(ii) $N(40n, 4n)$

(iii) 0.0207

(iv) 0.0456

Choice of limits is ± 2 standard deviations from the mean and so will include 95% of piles that contain 25 pamphlets.

18 (i) 0.037

(ii) 0.238

Assume that no time is lost during baton changeovers and that the runners' times are independent, i.e. that no runners are influenced by the performance of their team mates or competitors. The model does not seem entirely realistic in this.

19 (i) $N(2000, 1250)$

(ii) 1942

(iii) 0.7373

20 (i) 14%

(ii) 0.6

(iii) 15 m

(iv) 0.3043

21 (i) $S \sim N(600, 105.8)$; 0.0724

(ii) 0.839

(iii) 0.161

(iv) $\mu = 30.54$ g

22 (i) 0.127

(ii) $N(75m + 65f, 36m + 25f)$

(a) 0.0016

(b) 0

(iii) 0.1002

23 (i) 0.4258

Assume customers arrive randomly, singly and independently.

(ii) $N(372, 372)$; 0.8675

(iii) $N(30, 714)$; 0.13

24 (i) $f(x) = \dfrac{1}{2a}$ for $-a \leqslant x \leqslant a$

(ii) $0, \dfrac{a^2}{3}$

(iii) $0, \dfrac{na^2}{3}$

(iv) Greatest possible value $= \pm 5.0 \times 10^{-8}$.

Chapter 3

❷ (Page 79)

It is not possible to make a reliable judgement at this stage. A more detailed statistical investigation, using much more data, is necessary.

❷ (Page 81)

You would expect 90 out of the 100 to enclose 3.5.

Exercise 3A (Page 82)

1 (i) 5.205
(ii) 5.117, 5.293
2 (i) 47.7
(ii) 34.7 to 60.7
(iii) 27.3 to 68.1
3 (i) 0.9456 **(ii)** £7790–£8810
4 (i) (a) 0.1685 **(b)** 0.0207
(ii) 163.8–166.6
(iii) 385
5 78.4, 40.64;
76.44, 80.36;
The 95% confidence limits have been constructed from the sample data in such a way that 95% of similar-sized random samples will lead to an interval which will contain the true parent mean of the distribution.
0.2372
6 (i) 6.83, 3.04 **(ii)** 6.58, 7.08
7 (i) 5.71 to 7.49
(ii) It is more likely that the short manuscript was written in the early form of the language.
8 25.3, 3.6; 24.9, 25.8 (taking the last interval to be 38–40)
9

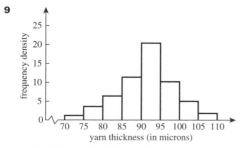

91.32, 7.41; 0.43; 90.5, 92.2

10 (i) Players' scores cannot be normally distributed because symmetry would require negative scores.
(ii) 11.79, 13.64
(iii) It would reduce the width of the confidence interval but the interval would be centred in the same place.
(iv) 409 000

❷ (Page 89)

Yes, but it would be worth mentioning that it is a marginal decision.

Exercise 3B (Page 89)

1 (i) 0.3085
(ii) 0.016
(iii) 0.0062
(iv) $H_0: \mu = 4.00$ g, $H_1: \mu > 4.00$ g
$z = 3$, significant
2 (i) $H_0: \mu = 72.7$ g, $H_1: \mu \neq 72.7$ g; Two-tail test.
(ii) $z = 1.84$, not significant
(iii) No, significant
3 (i) $H_0: \mu = 23.9°$, $H_1: \mu > 23.9°$
(ii) $z = 1.29$, significant
(iii) 4.42; This is much greater than 2.3 so the ecologist should be asking whether the temperature has become more variable.
4 (i) $H_0: \mu = 470$ days, $H_1: \mu > 470$ days
(ii) $z = 3.02$, significant
(iii) More time to produce offspring.
5 (i) $H_0: \mu = 80$ mph, $H_1: \mu \neq 80$ mph
$z = 2.28$, significant
(ii) Yes: $z = 1.33$, not significant
6 (i) $H_0: \mu = 14$ sea miles, $H_1: \mu < 14$ sea miles
(ii) $z = -2.284$, significant
(iii) Choosing 36 consecutive days to collect data is not a good idea because weather patterns will ensure that the data are not independent. A better sampling procedure would be to choose every tenth day. In this way the effects of weather patterns over the year would be eliminated.
7 $H_1: \mu \neq 1$ kg; $z = -1.79$, not significant
$H_1: \mu < 1$ kg; $z = -1.79$, significant.
8 $H_0: \mu = 50$ kg, $H_1: \mu < 50$ kg;
Yes: $z = -1.875$, significant
9 (i) 9900 g, 75.895 g, N(9900, 5760)
(ii) $110n, 64n$
(iii) 93

10 54; $z = -1.606$, not significant; 0.0288

11 (i) $N(190, 5.\dot{3})$

(ii) The skulls in group B have greater mean lengths and so a one-tail test is required.

(iii) 193.8

(iv) Significant

12 3.5, 2.917; 3.9, 2.49; 0.1208; 594

13 (i) $H_0: \mu = 43$ $H_1: \mu > 43$

(ii) Sample size large, $\overline{X}$ will have an approximately normal distribution.

(iii) Reject H_0 in favour of H_1.

(iv) $43.35 < \mu < 52.65$. This is consistent with part (iii).

❓ (Page 94)

More information is required. For example, if Phil Simmons had given the actual weights of the six bags he picked the standard deviation of the parent distribution could be estimated and use made of the t-distribution to answer the question.

Exercise 3C (Page 99)

1 0.01

(i) accept H_1

(ii) H_1

2 0.088

3 (i) $X < 59.82$

(ii) 0.052 to 0.057

4 (ii) 0.025

(iii) 0.0912

(iv) 0.9088

5 (ii) 0.24

(iii) 0.76

6 (i) $H_0: p = \frac{1}{2}$

$H_1: p > \frac{1}{2}$

where p is the probability of turning right

(ii) 27

(iii) 0.015

Chapter 4

Exercise 4A (Page 109)

1 (i) 322.9, 79.54

(ii) 278.8–366.9

2 (i) 66, 17.15

(ii) 51.7–80.3

(iii) Distribution of yield of all fruit farmer's trees is normal.

(iv) Number all the trees with different consecutive integers. Copy these integers onto separate pieces of paper; put these in a hat and pick out eight at random. The numbers chosen will identify the trees to be picked for the sample.

3 (i) 224.5–249.2

(ii) $H_0: \mu = 250\,g$, $H_1: \mu < 250\,g$

(iii) Assume the distribution is normal; $t = -1.914$, significant

(iv) The butcher's comment means that the other half were all underweight. This is much too big a proportion to be underweight. Furthermore, the underweight pies were often very much underweight.

4 (i) 18.25, 3.72

(ii) 16.32–20.18

(iii) The distribution of lengths of sentences written by the accused man is normal and the text represented by the sample sentences is representative of the general lengths of sentences he writes.

(iv) $H_0: \mu = 15.5$, $H_1: \mu \neq 15.5$

(v) $t = 2.56$, significant

(vi) No

5 (i) Monday

(ii) The 23 weekdays

(iii) 629.7–661.9

(iv) True, this distribution is not a normal one, but it may still be accurately modelled by one. $s = 37.24$ so step size is small compared with the standard deviation. It is very common in statistics to make a normal approximation to a discrete distribution and the results are usually very reliable.

6 (i) 63.6–72.0

(ii) Statistical: the distribution of tyre condemnation mileages is normal and the 12 tyres tested in the sample are representative of the distribution.

Practical: the tyres are tested under genuine working conditions.

(iii) $t = 3.047$, not significant

7 (i) $H_0: \mu = 750\,g$, $H_1: \mu < 750\,g$

(ii) $t = -4.57$, significant

(iii) The distribution of weights in the shoal is normal; this may be reasonable.

The sample is random; this is certainly not the case since all the pollack came from one shoal. The masses are independent; this may not be true when all the fish are taken from one shoal and so are likely to be of the same age.

8 (i) The distribution of total points in a hand is normal. This assumption is not fully justified because the distribution is not symmetrical about the mean; it is positively skewed. For example, you cannot have a hand with fewer than 0 points in it but you can get a hand with more than 20 points in it.

(ii) $H_0: \mu = 10$, $H_1: \mu < 10$

(iii) $t = -1.16$, not significant

9 (i) 89.8–93.3 (outlier excluded because something like an accident or a breakdown had clearly caused an atypical delay).

(ii) A one-sided confidence interval would result in the bus company claiming a shorter journey completion time than in the corresponding two-sided case. Also the confidence interval established in part (i) highlights the fact that the bus is very unlikely to complete the journey in less than 89.8 minutes. A one-sided confidence interval would not expose the company to such a lower-limit journey time.

(iii) $t = -4.498$, significant (if outlier excluded)

10 (i) 21

(ii) Distribution not normal

(iii) No because the confidence interval obtained would be too wide to be meaningful.

(iv) 45.0–66.3; The procedure will be valid provided that the distribution of life expectancies for this group is normal.

11 0.633–0.647

(i) 0.78, 0.160

(ii) 0.616–0.944

(iii) Large sample; no need for underlying normality

12 (i) $H_0: \mu = 110$, $H_1: \mu \neq 110$; accept H_0

13 (i) 172.7, 58.11; 166.95–178.45
The distribution of 17-year-old boys' heights is normal and the sample used in the experiment is random.

(ii) A larger sample will reduce the width of the confidence interval for μ because (1) the standard deviation of the sample means is $\dfrac{\sigma}{\sqrt{n}}$ and so, as n increases, this will decrease

(2) the value of k in $\dfrac{k\sigma}{\sqrt{n}}$ is less.

(iii) Normality no longer required; randomness still needed.

14 (i) Mean = 2.36, standard deviation = 2.74; $a = -0.558$, $b = 5.278$
Assumptions: journey times are a random sample; distribution of T is normal.

(ii) Commuter's interpretation is wrong. It should be '90% of such confidence intervals should contain the true mean'.

(iii) Accept H_0 that $\mu = 0$, i.e. refute commuter's suspicions at 5% level.

(iv) Confidence interval is likely to be narrower because of lower percentage-point for t-value and lower standard error since sample size is large.

15 (i)

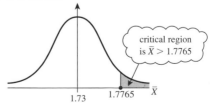

(ii) $0.616 < \mu < 0.804$
It is assumed that (1) the throws are independent, (2) that the sample size is large enough for the estimate of σ^2 to be reliable enough to regard σ^2 as known and (3) that the sample size is large enough for the central limit theorem to indicate that $\bar{x}$ is normally distributed.

(iii) 177. The standard deviation of the parent distribution is unknown and the distribution of R is not normal.

(iv) The population is not normal and the sample is small, therefore, neither the t-distribution nor the central limit theorem can be used.

16 (i) The method must ensure each club member has an equal chance of being chosen. Could put names in a hat and select at random or use the random number generator on a calculator.

(ii) mean = £4.74, $s^2 = 2.74$

(iii) $3.56 < \mu < 5.92$.
Background population is normal.

(iv) By increasing sample size; 13

17 (i) $H_0: \mu = 1.73$, $H_1: \mu > 1.73$

(ii) $\overline{X} \sim N\left(1.73, \dfrac{0.08^2}{8}\right)$

(iii) $\overline{X} > 1.7765$

(iv) Accept H_0; assume eight players are a random sample of men who play basketball.

(v) 0.134

18 1.934–2.052, 9.373–10.551; 120, halved

Chapter 5

? (Page 121)

The data are real. Any model is just your theory.

? (Page 129)

The fit looks suspiciously good but see text that follows.

Exercise 5A (Page 132)

Note: in some of the questions in this exercise you have to combine different classes. There may be some variation in the value of X^2 according to the way you have done this.

1 (i) H_0: The number of mistakes on a page can be modelled by the Poisson distribution.

H_1: The number of mistakes on a page cannot be modelled by the Poisson distribution.

$X^2 = 36.3$, significant

(ii) The mean rate may not be constant, for example, she may make more mistakes when she is tired. The mistakes might not be independent if, for example, some sections are about things she cannot spell.

2 (i) Binomial, $B(5, \frac{1}{4})$

(ii) H_0: The number of white flowers in each tray can be modelled by a binomial distribution, $B(5, \frac{1}{4})$.

H_1: The number of white flowers in each tray cannot be modelled by this distribution.

$X^2 = 0.343$, not significant

3 (i) H_0: The two dice used in the casino are fair.

H_1: The two dice used in the casino are not fair.

$X^2 = 10.1$, not significant

4 H_0: The students guessed the answers at random.

H_1: The students did not guess the answers at random.

$X^2 = 18.8$, significant

5 H_0: In 1999 the same proportion of candidates get the various grades as in the past.

H_1: In 1999 a different proportion of candidates get the various grades.

$X^2 = 1.818$, not significant

6 (i) H_0: The size of rocks is distributed evenly on the scree slope.

H_1: The size of rocks is not distributed evenly on the scree slope.

$X^2 = 7.82$, significant

(ii) The test endorses the theory.

7 H_0: The roulette wheel is fair.

H_1: The roulette wheel is biased.

$X^2 = 8.63$, not significant

8 (i) $\bar{x} = 0.933$, $sd^2 = 0.929$

(ii) H_0: These data can be modelled by the binomial distribution.

H_1: These data cannot be modelled by the binomial distribution.

$X^2 = 0.28$, not significant

(iii) Poisson, because of general spread of data in table and because mean $\approx$ variance.

9 (i) Several observed frequencies are too small. In order to have $f_e \geqslant 5$ in each class there would be only two classes. There are two constraints and so no degrees of freedom, therefore the χ^2 test cannot be used.

(ii) H_0: The occurrence of Morag's spelling mistakes may be modelled by the Poisson distribution.

H_1: The occurrence of Morag's spelling mistakes may not be modelled by the Poisson distribution.

$X^2 = 2.46$, not significant

(iii) Spelling mistakes occur singly, randomly and independently. This could be realistic.

10 H_0: These data can be modelled by the Poisson distribution.

H_1: These data cannot be modelled by the Poisson distribution.

$X^2 = 7.25$, not significant at 5% level

11 (i) H_0: The new drink is indistinguishable.

H_1: The new drink is distinguishable.

$X^2 = 5$, significant at 5% level

(ii) $0.0207 < 0.05$, significant at 5% level (one-tail test); both give the same answer

(iii) Such a small sample is highly unlikely to provide the company with information it can trust.

12 (i) Reject the null hypothesis because the test statistic is much larger than the critical value, 9.488, at the 5% significance level.

(ii) $X^2 = 0.57$, not significant

13 (i)

Number of children	0	1	2	3	4	5+
f_e		246.6	345.2	241.7	112.8	39.5	14.2

(ii) H_0: Number of children per household can be modelled by the Poisson distribution with $\lambda = 1.40$.

H_1: Number of children per household cannot be modelled by the Poisson distribution with $\lambda = 1.40$.

$X^2 = 32.17$, significant

14 (i) H_0: Shoplifting is equally likely to occur in all months.

H_1: Shoplifting is more likely to occur in some months than others.

$X^2 = 14.28$, not significant at 5% level.

(ii) $$X^2 = \sum \frac{(f_o - f_e)^2}{f_e} = \sum \frac{f_o^2 - 2f_o f_e + f_e^2}{f_e}$$

$$= \frac{1}{f_e} \Sigma f_o^2 - 2\Sigma f_o + \Sigma f_e$$

$$= \frac{1}{f_e} \Sigma f_o^2 - \Sigma f_o \quad \text{because} \quad \Sigma f_o = \Sigma f_e$$

15 (i) 2.5

(ii)

x	1	2	3	4	5	6	7	8+
f_e	80.0	48.0	28.8	17.3	10.4	6.2	3.7	5.6

(iii) H_0: Number of attempts follows model suggested.

H_1: Number of attempts does not follow model suggested.

Conclusion: $10.36 < 11.07$ so accept H_0

Number of degrees of freedom $= 8 - 3 = 5$

i.e. number of classes $- 1$ for total $- 1$ for using sample mean and -1 for combining last two classes.

16 (i)

	A	B	C	D	E	F	G	H
f_e	71	113	104	71.5	57	49	29.5	5
f_o	53	105	111	80	63	49	36	3

(ii) H_0: Distribution of properties in Trumpton reflects the national distribution.

H_1: Distribution of properties in Trumpton does not reflect the national distribution.

Conclusion: $9.475 < 12.02$ so accept H_0

(iii) There is now enough evidence to reject H_0.

17 (i)

n	0	1	2	3	4 or more
f_e	9.47	18.95	18.95	12.63	10

(ii) $\chi^2 = 2.006$, accept H_0

18 (i) f_e: 13.65   18.14   22.43   18.14   13.65

(ii) 2.56; accept model as in part (i).

Index